W9-AMB-123

*We're in This War, Too*

# We're in This War, Too

# WORLD WAR II LETTERS FROM
# AMERICAN WOMEN IN UNIFORM

*Judy Barrett Litoff*
*David C. Smith*

*New York    Oxford*
*Oxford University Press*
*1994*

Oxford University Press

Oxford  New York  Toronto
Delhi  Bombay  Calcutta  Madras  Karachi
Kuala Lumpur  Singapore  Hong Kong  Tokyo
Nairobi  Dar es Salaam  Cape Town
Melbourne  Auckland

and associated companies in
Berlin  Ibadan

Copyright © 1994 by Judy Barrett Litoff & David C. Smith

Published by Oxford University Press, Inc.,
200 Madison Avenue, New York, New York 10016

Oxford is a registered trademark of Oxford University Press

Library of Congress Cataloging-in-Publication Data
We're in this war, too : World War II letters from American women in
uniform / [edited by] Judy Barrett Litoff, David C. Smith.
p.  cm.  Includes bibliographical references and index.
ISBN 0-19-507504-8
1. World War, 1939–1945—Personal narratives, American.  2. Women
soldiers—United States—Correspondence.  I. Litoff, Judy Barrett.
II. Smith, David C. (David Clayton), 1929–        III. Title: We're
in this war, too.
D811.W427  1994
940.54'8173—dc20  93–36523

9 8 7 6 5 4 3 2 1

Printed in the United States of America
on acid free paper

*For*
*James Harvey Young*
*and*
*Paul Wallace Gates*
*who*
*transcended . . .*

# Acknowledgments

Over the past decade, many friends and colleagues throughout the United States and Europe have provided support for our search for "the missing letters" written by United States women during World War II. We are hesitant to mention them by name, for it is unlikely that our list will be all-inclusive. Yet we still feel compelled to single out particular individuals whose help and expertise have proven to be of special importance.

At Bryant College, we would like to offer thanks to Bill Hill, Mary Lyons, Hinda Pollard, Steve Frazier, Bob Sloss, Connie Cameron, Gretchen McLaughlin, Colleen Anderson, Patricia Sinman, Bob Di- Prete, and Conny Sawyer. Thanks is also due to the many student assistants who have helped with the project, including Maria Acamp- ora, Todd Balcom, Tracy Banasieski, Scott Byrne, Tedd Dutter, Gretchen Golembewski, Meredith King, and Joanna Powers. The stu- dents who enrolled in the honors seminar, U.S. Women and World War II, went "above and beyond the call of duty." Graduate assistant Birgit Neumann, a former Army staff sergeant, added immeasurably to this work. At the University of Maine, we wish to thank Jerry Nadel- haft, Howard Segal, Ed Schriver, Marli Weiner, Richard Blanke, Bill Baker, Frank Wihbey, Mel Johnson, Dawn Lacadie, Suzanne Moulton, and Debbie Grant. Graduate student Kim Sebold took time out from her own work to locate important documents for us. Special thanks is due the members of History 499, a special seminar on women and World War II. Moreover, our academic institutions have assisted in our work by awarding us course releases, summer research grants, travel funds, and support for photocopying and other clerical ex- penses.

We are fortunate to have a number of colleagues who have gener- ously shared their expertise and knowledge with us. They include Re- gina Akers, Leah Atkins, Judy Austin, Bob Baron, Judith Bellafaire, Alison Blunt, Malcolm Call, Dave Danbom, Dottie DeMoss, David Demeritt, Lowell Dyson, Jamie Eves, Anne Effland, Mary Giunta, Alfred deGrazia, Henry Gwaizda, Susan Hartmann, Don and Peggy Hoffman, John Inscoe, Penelope Krosch, Larry Malley, Gerry McCauley, Gordon Marsden, Nina Mjagkij, Lynn Parsons, Bill Pratt, Wayne Rasmussen, Glenda Riley, Gillian Rose, Fred Schmidt, Sharon Seager, Janet Sims-Wood, Peter Soderbergh, Martha Swain, Bill Tuttle, Detlef Vogel, and Sarah Wilkerson-Freeman.

Two historians who epitomize the meaning of the words friend and colleague are Stephen Ambrose and D'Ann Campbell. We are especially appreciative of their contributions to our understanding of the larger meaning of World War II as well as their unfailing support of our efforts to reconstruct the histories of "ordinary" women during wartime.

Archivists, librarians, and museum directors around the nation have helped to locate obscure wartime letter collections. In particular, we would like to thank the following for their assistance: Debra L. Anderson, University of Wisconsin-Green Bay; Ann E. Billesbach, Nebraska State Historical Society; D'Ann Blanton, National Archives; Thomas W. Branigar, Dwight D. Eisenhower Library; Jerry G. Burgess, The Women's Army Corps Museum; Bernard F. Cavalcante, Naval Historical Center; Fred Daugherty, 45th Infantry Division Museum; Carolyn Dilulio, U.S. Coast Guard Museum; Jeff Flannery, Manuscript Division, Library of Congress; Anne Frantilla, Bentley Historical Library, University of Michigan; Diane Gutscher, Bowdoin College Library; James J. Holmberg, The Filson Club Historical Society; Johanna Koehn, U.S. Army Medical Department Museum; Dawn Letson, Blagg-Huey Library, Texas Woman's University; Lt. Col. James R. McLean, U.S. Army Military History Institute; David Ment, Milbank Memorial Library, Teachers College, Columbia University; Mike Miller, Marine Corps Historical Center; Barbara Pathe, American Red Cross National Headquarters; Duane J. Reed, U.S.A.F. Academy Library; Colleen Schiavone, Kornhauser Health Sciences Library, University of Louisville; William D. Welge, Oklahoma Historical Society; and, Lt. Col. Iris J. West, Army Nurse Corps, U.S. Army Center of Military History.

As we have traveled around the nation speaking to audiences about our women and letters project, we have met many enthusiastic individuals who have contributed to our work, and they have done so in a variety of ways. They include Prudence Burns Burrell, Denver Gray, Ruth M. Lee, Madge Rutherford Minton, Frances Robinson Mitchell, Dagmar Noll, Yvonne "Pat" Patemann, Oscar Rexford, Marianne Verges, June Wandrey, Elbert Watson, and Claudine Zanella.

At Oxford, we are indebted to Sheldon Meyer, the doyen of editors, for shepherding us through this project. We are most appreciative of the careful attention that he and his staff, especially Joellyn Ausanka and Karen Wolny, have given to this book.

Of course, *We're in This War, Too* could not have been written had it not been for the hundreds of World War II letter writers from throughout the United States who readily shared their wartime correspondence with us. Indeed, one of the most difficult challenges we faced in preparing this book was choosing the letters to be published. For every letter that is printed, there are dozens of others that could have been substituted, as only a small fraction of the thousands of poignant and powerful letters we have collected appear in *We're in This War, Too*. We are grateful to each and every person who contributed letters to us—especially those individuals whose letters do not appear in this book.

Family and close friends know more about women's correspondence from World War II than they ever thought possible. We appreciate this "captive" audience and offer our love and thanks to Dorothy W. Barrett, Hal Borns, Joshua Eves, Kit Smith Eves, Alyssa Barrett Litoff, Hal Litoff, Nadja Barrett Litoff, Renee and David Califf, Wayne Seabolt, Clayton Smith, Sylvia Smith, Barbara Wooddall Taylor, and Robert Wooddall.

J. B. L.
D. C. S.

# Contents

Introduction: The Case of the Missing Letters, 3

1. The Coming of War, 11

2. A Woman's Place Is in Uniform, 29

3. Stateside—Zone of the Interior, 81

4. The War Against the Axis: Italy and Germany, 121

5. The War Against the Axis: Japan, 177

6. Preparing for the Postwar World, 221

For Further Reading, 261

Index, 265

*We're in This War, Too*

You ask if I'd like to go home. More than anything on earth, but I couldn't leave now. Our work is just beginning.

—LT. AILEEN HOGAN,
Army Nurse Corps
Somewhere in France
Christmas Day, 1944

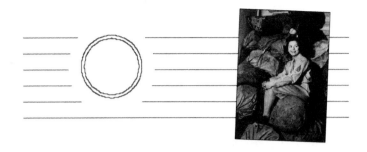

# Introduction: The Case of the Missing Letters

FOR THE PAST DECADE, we have been involved in a nationwide search to locate letters written by United States women during World War II. In the early 1980s, when we began this search, many of our colleagues and friends discouraged us from taking on this challenge because of the common wisdom that few, if any, letters written by American women had survived the vicissitudes of the war and the postwar years. After all, men in combat were under orders not to keep personal materials such as diaries and letters. Moreover, we were repeatedly warned that, should we locate letters written by women in uniform, they would contain little if any information of significance because of censorship regulations. Today, some 30,000 letters and ten years later, we can state unequivocally that the common wisdom was wrong.

Our search for women's wartime correspondence has led us down many circuitous paths and avenues. It began in 1980 as we conducted the research for a book, *Miss You: The World War II Letters of Barbara Wooddall Taylor and Charles E. Taylor* (1990), which was based on thousands of pages of correspondence between a young war bride and her soldier husband. The letters of Barbara Wooddall Taylor provided us with so much valuable, and heretofore largely unfamiliar, information about the experiences of a typical young woman who "grew up" during the wartime years that we decided to seek out other letters written by United States women between 1940 and 1946.

Following an extensive search of the standard literature on World War II, it became apparent to us that efforts to identify and locate women's wartime correspondence had escaped the attention of scholars. Thus, in the spring of 1988, we decided to devise a brief author's query requesting information from anyone who had knowledge about letters written by United States women during the Second World War. We sent the query to every daily newspaper in the United States—about 1,500 newspapers in all—and requested that it be printed on the letters-to-the-editor page.

Much to our delight, newspapers throughout the United States published our query. Very shortly thereafter, wartime letters from across the United States began to pour into our offices. By midsummer 1988, we realized that we had struck a gold mine of information.

We supplemented our author's query to the nation's newspapers with more than 500 letters of inquiry to magazines and newsletters specializing in issues of concern to women, veterans, and minorities. We wrote letters about our search to every state historical society and to dozens of research and university libraries. In an effort to locate the correspondence of African-American women, we solicited the advice of prominent black historians, surveyed archives specializing in African-American history, and even sent out a special appeal to 500 predominately black churches around the nation. In total, we have written more than 2,500 letters of inquiry. We often wonder if this might qualify us for inclusion in the *Guinness Book of World Records*.

Our search for the "missing letters" has taken us to twenty-three states and Washington, D.C. In fact, our travels in search of women's wartime correspondence probably deserve an article, if not a book, of their own. We have aired appeals on radio and television stations across the nation, and we have written feature articles about our project for newspapers and magazines. We rarely turn down invitations to speak to interested groups about the project. As a result, our audiences have ranged in size from two to one thousand persons. One thing we have learned from this effort is that we can never predict when we will locate a new bonanza of letters.

We send out a "V-Mail Update" twice a year to the more than 800 people throughout the United States who have expressed an interest in our work. Not only does this update enable us to keep in touch with everyone on our mailing list, but it also serves to disseminate information about the scope of the project. Moreover, many of the recipients of the "V-Mail Update" have been kind enough to share it with others who have, in turn, contributed letters to the project.

In the process of collecting the letters in our archive, we have learned that in order to obtain letters, we ourselves must write letters.

While we have never taken the time to count the number of followup letters we have written, the extent of this correspondence is enormous.

The 30,000 letters we have collected were written by more than 1,500 women representing diverse social, economic, ethnic, and geographic circumstances. We have collected letters written by grade-school dropouts, but we also have letters penned by college graduates. Our archive includes letters written by women from rural and small-town America as well as large metropolitan areas. The letters of mothers, daughters, sisters, cousins, aunts, grandmothers, stepmothers, wives, sweethearts, and friends all form a part of our archive.

During the early stages of this project, we focused our attention on reading, analyzing, and publishing information about the home front letters of civilian women in the United States. This effort resulted in two books, *Since You Went Away: World War II Letters from American Women on the Home Front,* and *Dear Boys: World War II Letters from a Woman Back Home,* both of which were published in 1991. Yet, even before the page proofs of these books had been corrected, we had already begun the groundwork for a new book on the wartime letters of United States women in uniform.

Among the initial surge of letters we acquired in the spring and summer of 1988 were letters written by women who served in the Women's Army Corps (WAC), the Women's Reserve of the Navy (WAVES), and the American Red Cross (ARC) during World War II. With these letters as a nucleus, we began to search for the wartime correspondence of Army and Navy nurses, the Marine Corps Women's Reserve, the Coast Guard Women's Reserve (SPAR), the Women Airforce Service Pilots (WASP), and civilians in uniforms. At the same time, we also endeavored to augment our collections of letters written by WACs, WAVES, and ARC women.

We sent out a new set of inquiries to the various women's veterans' organizations as well as to all known, living black Army nurses. We visited a number of military archives, including the Women's Army Corps Museum, the United States Coast Guard Museum, the Marine Corps Historical Center, and the United States Army Center of Military History. Cooperative librarians at the United States Army Military History Institute, the Naval Historical Center, the American Red Cross National Headquarters, the Army Medical Department Museum, the United States Air Force Academy Library, and many other archives photocopied reams of letters for us. We also discovered relevant letter collections at the National Archives and the Manuscript Division of the Library of Congress. Still, the vast majority of the letters that appear in this book are ones that were donated to us by "ordinary" individuals who simply heard of our appeal.

Many of the women who have donated letters to our archive have included the caveat that they doubt that there is anything of value in their missives because they were careful to follow the dictates of strict wartime censorship regulations. Others have apologized for the cheery, upbeat quality of their letters, noting that they did not want to cause the recipients, who were often family members, undue worry and stress. These same letter collections, however, contain eyewitness accounts from women who dodged "buzz bombs" in England, helped perform emergency surgery in evacuation hospitals at the front, distributed coffee and doughnuts to men in battle, established Red Cross clubs in remote areas around the globe, provided aid and comfort for returning prisoners of war who had been incarcerated at Japanese internment camps, and helped care for the survivors of Dachau. Clearly, censorship regulations were not nearly so strict as many have believed. In fact, what is most extraordinary about the letters in our archive is how much—rather than how little—frank and detailed commentary they contain.

Women's wartime letters offer readers the opportunity to view a dimension of women's past that has often been disregarded and discounted. *We're in This War, Too* has been written to redress this oversight. This is not a book based on what men said and wrote about wartime women. Nor is this a book of contemporary reflections tempered by a half-century of successive events. Rather, these letters are honest accounts written "at the scene" for a limited audience and with little idea that historians would one day be interested in their content. They offer perceptive insights into heretofore unexplored, but fundamental, aspects of the war.

The letters include telling accounts of the courage of African-American women as they combatted racism at home and fascism abroad; the agony and isolation experienced by the only Jewish servicewoman at her duty station; glimpses of the stress and strain that lesbians in the military encountered; the blossoming of heterosexual love in the face of battle; coping with the tragedies of war's carnage; and the intense camaraderie shared by women in uniform as they assumed new and challenging responsibilities in behalf of the war effort. Finally, they contain discerning commentary about women's hopes and dreams for the postwar world as they planned for reunions with loved ones and grappled with questions concerning marriage, family, education, and careers.

In short, the letters of these "pioneers" provide clear and unequivocal evidence of the many important ways that women actively participated in the war effort, and they vividly illustrate women's growing sense of self and their place in the world. Our reading of these letters has convinced us that women's wartime correspondence captures the complexity and essence of the experience of war for women better than any available source.

As we launched our search for the letters of women in uniform, we knew that we wanted to collect letters that represented the spectrum of experiences of "ordinary" women, for it is the voices of "ordinary" citizens that have traditionally remained hidden from the pages of history. Thus, with one exception, letters written by the directors of the WAC, WAVES, SPARs, Women Marines, Army Nurse Corps, Navy Nurse Corps, and WASP are not included in this volume. War correspondents usually wore military uniforms without military insignia, as did some USO entertainers. But we did not seek out the letters of luminaries such as Margaret Bourke-White or Martha Raye. Instead, this book is based on the letters of the rank and file—enlisted women, junior grade officers, Army and Navy nurses near the front lines of battle, Red Cross clubmobile workers, and many other forgotten heroines of World War II.

The letter writers in this book represent a mature, well-educated, segment of the population. The minimum age requirement for each of the women's services was twenty. For Red Cross women who served overseas, the minimum age was twenty-five. All of the service branches preferred that women have high school educations. Officers were expected to be college graduates or to have two years of college education plus two years of work experience. American Red Cross workers were also expected to have some college education. Because Army and Navy nurses were required to be graduates of accredited nursing schools, they were usually in their mid-twenties or older when they entered the service. Members of the Women Airforce Service Pilots, who were required to have a pilot's license before being accepted into the program, were also both well-educated and mature.

Modern readers may be surprised and jarred by the fact that these mature, well-educated women, who repeatedly acknowledged the importance of the camaraderie they shared with other women in uniform, regularly referred to themselves and their co-workers as "girls" and "gals" and to the men at the battle fronts as "boys." We want to remind readers that to hold women of the 1940s to the linguistic standards of the 1990s is both inappropriate and incorrect.

Except for the occasional addition of commas and periods to ease readability, we have presented the letters exactly as they were written. We have used ellipses to indicate when material has been deleted and brackets to denote when we have added explanatory information.

Despite the far-reaching scope of our search for women's wartime correspondence, it is not possible to determine whether the numerous and diverse letters collections in our archive are representative of the billions of letters penned by United States women during World War II. Yet based on what Linda Gordon has described as the "methodological principle of saturation," we are reasonably certain that we have tracked down a substantial sampling of the variety of letters written by United States women during the war years.[1] Historians have always been required to evaluate their sources and to piece together the meaning of the past from fragmentary bits of information. Even with the new and sophisticated cliometric approaches to the study of the past, the writing of history, ultimately, depends upon qualitative analysis. In making the selections for this book, we have chosen those letters which, in our best judgment, represent the breadth and richness of the wartime experiences of United States women in uniform.

The case of the missing letters is not closed. We continue to search for new letter collections to include in the forthcoming microfilm edition of our entire archive of letters. We are especially interested in locating additional letters from women who were stationed in Alaska, the Caribbean, and Burma. While the practice of segregation and discrimination created a wide array of difficult problems for African-American women in uniform, the letters they wrote form a crucial part of our nation's past. Nonetheless, despite our best efforts, letters from African-American women, as well as other minority women, are underrepresented in our archive. We also need letters from Cadet Nurses, members of the Women's Land Army, the American Women's Voluntary Services, and the Salvation Army. We are still searching for letters written by lesser-known war correspondents and USO entertainers. As far as we are concerned, the case of the missing letters will never be closed.

As the letters in this book demonstrate, women in uniform joined

---

1. Linda Gordon, "Black and White Visions of Welfare: Women's Welfare Activism, 1890–1945," *Journal of American History* 78 (September 1991):562–63; and Linda Gordon, "Social Insurance and Public Assistance: The Influence of Gender in Welfare Thought in the United States, 1890–1935," *American Historical Review* 97 (February 1992):22.

in the nation's battle against fascism with energy, courage, fortitude, wit, and ingenuity. For many, World War II was the defining event of their lives. A recurring theme in these letters is how the experience of war afforded uniformed women the opportunity to become stronger, more capable, and more tolerant individuals who were better equipped to meet the challenges of the postwar world. Historians have devoted considerable attention to the question of what happened after the war, and significant works on this topic have been published.[2] Yet before we can comprehend the full import and complexity of the history of American women in the postwar world, we must first arrive at a clearer understanding of how women themselves perceived and experienced the World War II era. *We're in This War, Too* is one contribution to this task.

---

2. For examples, see the following: Susan Hartmann, "Prescriptions for Penelope: Literature on Women's Obligations to Returning World War II Veterans," *Women's Studies* 5 (1978):233–39; Susan M. Hartmann, *The Home Front and Beyond: American Women in the 1940s* (Boston: Twayne Publishers, 1982); D'Ann Campbell, *Women at War with America: Private Lives in a Patriotic Era* (Cambridge, Mass.: Harvard University Press, 1984); Elaine Tyler May, *Homeward Bound: American Families in the Cold War Era* (New York: Basic Books, 1988); William H. Chafe, *The Paradox of Change: American Women in the Twentieth Century* (New York: Oxford University Press, 1991); Joanne Meyerowitz, "Beyond the Feminine Mystique: A Reassessment of Postwar Mass Culture, 1946–1958," *Journal of American History* 79 (March 1993):1,455–82.

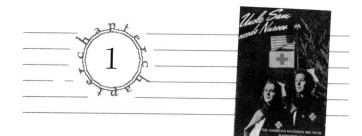

# The Coming of War

T HE JAPANESE ATTACK ON Pearl Harbor precipitated the formal entry of the United States into World War II. Yet many
American women and men had taken part in the nation's preparedness
campaign to defeat fascism long before the events of December 7,
1941. American citizens fought in the Spanish Civil War during the
mid-1930s. American Red Cross (ARC) volunteers came to the aid of
the Chinese in their struggle against Japanese invasion forces as early
as 1937. By the time of the London Blitz in the late summer of 1940,
American volunteers were serving as nurses in British hospitals, flying
Spitfires and Hurricanes for the Royal Air Force, and shipping "Bun-
dles for Britain" to their beleaguered friends across the Atlantic.

Recognizing the inevitability of America's entrance into the conflict, the United States Congress enacted the Selective Service and
Training Act, the first peacetime draft in the nation's history, in September 1940. On October 29, with the drawing of the first draft numbers, the reality of war was brought home to many Americans. Over
the course of the next twelve months, a succession of events pushed
the nation steadily toward war. In March 1941, Congress passed the
Lend Lease Act. During the latter part of May, the Office of Civilian
Defense was created, and a week later President Franklin D. Roosevelt
declared a state of national emergency. In August 1941, the ties con-
necting the United States to Great Britain were further cemented with
the announcement by the two countries of the Atlantic Charter "to
ensure life, liberty, independence and religious freedom and to pre-

serve the rights of man and justice." By September, an "undeclared" shooting war with German submarines had broken out.

During the summer of 1941, the uneasy relationship between the United States and Japan also deteriorated with the tightening of the trade embargo and the freezing of Japanese assets. Even the arrival of special Japanese envoys in Washington on November 15 did little to alleviate the worsening situation between the two countries. The events of December 7, 1941, in Hawaii and the western Pacific outraged the American citizenry. Nonetheless, by the time of the Pearl Harbor attack, the entry of the United States into the Second World War had been anticipated for many months.

WHEN THE UNITED STATES entered World War II, the opportunities for women to serve in the military were very limited. In fact, only registered nurses qualified for military service. United States women had first distinguished themselves as combat nurses during the American Civil War. Following the formation of the Army Nurse Corps (ANC) in 1901 and the Navy Nurse Corps (NNC) in 1908, more than 20,000 women served as military nurses both at home and around the globe during World War I. At the time of Pearl Harbor, the Army Nurse Corps numbered 5,433 members, while 823 women were members of the Navy Nurse Corps. By the end of World War II, with the help of American Red Cross recruitment campaigns, the number of military nurses had increased tenfold, with 54,291 women serving in the Army Nurse Corps and 11,086 women serving in the Navy Nurse Corps. All told, 76,000 women, representing 31.3 percent of all active professional nurses, served in the armed forces during World War II.[1]

Because of their professional training, women who entered the Army and Navy Nurse Corps were treated as officers. However, at the beginning of the war, they were awarded "relative rank" as commissioned officers in which they received the pay of officers but not full command authority. In 1944, nurses in both services were given full officer status.

With the departure of so many registered nurses from civilian to

---

1. Philip A. Kalisch and Beatrice J. Kalisch, *The Advance of American Nursing* (Boston: Little, Brown and Company, 1978), 485; and "The Nurses' Contribution to American Victory," *American Journal of Nursing* 45 (September 1945):683–86.

military assignments, the United States soon faced a shortage of civilian nurses. In June 1943, in response to this shortage, Congress adopted the Bolton Act, creating the Cadet Nurse Corps training program. The Bolton Act, named after its chief sponsor, Representative Frances P. Bolton from Ohio, provided federal funds for nursing students who promised to perform essential civilian or military nursing duties upon graduation. This proved to be a tremendously successful program.

Even with the Bolton Act, however, a shortage of military nurses developed late in 1944 as battle casualties mounted. In his January 1945 State of the Union address, President Roosevelt called for the drafting of nurses into the armed forces. Such a bill was passed by the House of Representatives early in 1945, but it was still under consideration in the Senate when the war in Europe ended in May 1945. By that time, enough nurses had volunteered to meet the anticipated needs of the war in the Pacific, and no further action was taken on the bill to draft nurses.

Army and Navy nurses stationed in Hawaii during the months prior to Pearl Harbor wrote perceptive letters to Stateside relatives and friends in which they described the buildup to the coming of the war as well as the rising tensions in the Pacific. Blanche M. "Rusty" Kiernan of Denver, Colorado, joined the Army Nurse Corps in 1939. In March 1941, she was transferred to Hawaii, where she was stationed at Tripler General Hospital. Her weekly letters to her mother vividly depict life in Hawaii in the months before the United States entered the war. (For other letters by Kiernan, see pages 179–180.)

Honolulu, Territory of Hawaii [T.H.]

Dearest Mom, March 25, 1941

Well, here I am and I still can't believe it. . . . The good ship *Etolin* [Kiernan's troop ship] sailed promptly at high noon and I had the strangest feeling when we sailed under the Golden Gate bridge and the shore line melted away. It seemed then as though two years would be an eternity but the girls here say that the time flows by very quickly. . . . The quarters here are quite nice but they are crowded so for the present my bed is in an upstairs hall, but I will have a room soon as several of the girls who have been here longer are moving out. . . . I forgot to tell you that on our second afternoon out a fire and boat drill was held. Everyone, even the little children, had to put on a life belt or preserver and report to their inspection stations at a given signal. . . . My trunk was delivered today so I guess I am set for the two years. I had a couple of hours yesterday when I felt like the man without a country. . . . Be good, darling and take good care of yourself and please write often to this "Malihini" (newcomer).

Love and kisses, Blanche


Dearest Mom and C[onnie],

Hickam Field, [T.H.]
March 28, 1941

. . . Arrangements have been made for eight of us to have quarters at Hickam Field and ride back and forth to duty in an ambulance. . . . The quarters are brand new and it is a duplex. . . . They are building a 160 bed hospital at Hickam and plan to open it in October and assign nurses for duty there, so perhaps we will have first choice. . . .

Love and Kisses, Blanche


Darling Mom,

[Hickam Field, T.H.
April 6, 1941]

. . . Black-outs and alerts are practiced quite often and before long there will be a general black-out rehearsal for all the islands. During alerts everyone remains at his station but the lights are left on. There is also a possibility that the island of Oahu, where we are, will stock up enough food and supplies for six months ahead. . . . The influx of army and navy personnel is so great that the buildings in some of the posts aren't sufficient to care for them so many officers and families have to live on commutation. . . . I think within a year though there will be many quarters on the posts. . . . The U.S.A. transport *Republic* stopped here Monday on its way from Frisco to the Philippines. It docked at 7:00 a.m. and sailed at 6:30 p.m. There were over 2,000 troops aboard. . . .

Love and kisses, B.


Hello, my Darlings,

[Hickam Field, T.H.]
May 22, 1941

. . . Today I got up early and went . . . to Ford Island [in the middle of the Pearl Harbor lagoon] to visit a friend . . . whose husband is a lieutenant in the Navy. . . . Their house is on the edge of the island and they practically have two destroyers in their front yard. [He] was on maneuvers so I didn't get to meet him. He is a pilot and has been flying almost constantly for over a week. . . . The blackout Tuesday night was a huge success and very interesting as everyone cooperated. At 9:00 p.m. there was a plane over each island broadcasting what they saw. Finally the flash came that enemy planes were approaching Oahu and the blackout signal was given with sirens. . . . The lights remained off for about twenty minutes and we could hear planes overhead. . . . The other morning 21 B-17 4 motor flying fortresses [bomber planes] flew in from Frisco and landed at Hickam Field. . . .

Love and Kisses, Rusty

Hi, Mom and Sis,                                      [Hickam Field, T.H.]
                                                      July 28, 1941

. . . We are on an alert now and if we venture away from quarters we must leave a phone number. Any night now we expect to be called out to Tripler to report at our post. All of the army posts here are on an alert at present and we don't know just how long it will last. I understand there isn't as much war talk here as there is at home, but we have no fear because this island is so well fortified no enemy could get within miles of it. There are guns all over the place and planes go out to sea on scouting missions every day. I really worry more about you than you should about me. . . . All of my corpsmen are being taken off the wards August 1st and sent into the field. They will be replaced by civilian Filipinos. . . . That will really be a headache, but the good old A.N.C. will carry on. . . .

Love and kisses, Rusty (the Hula kid)

Dear Madames,                                      [Hickam Field, T.H.
                                                   September 11, 1941]

. . . . I just finished listening to the President's speech for the second time [Kiernan is referring to FDR's "shoot first order" speech, which marked the beginning of the undeclared naval war] and he sounds like he means business. . . .

I love you both and miss you terribly, Rusty

Mom Darling,                                       [Hickam Field, T.H.]
                                                   October 26, 1941

Before I turn in for the day I must write a few lines so you won't worry about me. . . . All the nurses living off the post have to move in to Tripler by the 1st except the 8 of us in our quarters here at Hickam. . . .

Your big baby, Rusty

[Hickam Field, T.H.
November 2, 1941]

Dearest Mom,

. . . I also received a clipper letter [via the transpacific airmail service begun by Pan American Airways in 1934] from Walt yesterday and he said you were worried because you had not heard from me for two weeks. Please don't worry any more, Mom, because the mail schedules are so uncertain now. . . .

Love and Kisses, Rusty

[Hickam Field, T.H.]
November 6, 1941

Hello Momsie,

. . . Tuesday, I had the good fortune to go aboard the *USS Solace* [the crew of the hospital ship *Solace* played a crucial role in caring for the wounded on December 7, 1941] which is the navy hospital ship which follows the fleet in the Pacific. It is anchored in Pearl Harbor and I went with one of the nurses from Tripler who was a classmate of one of the nurses on board. . . . The ship is equipped to care for about 450 patients and has three operating rooms, a pharmacy, laboratory, physiotherapy dept., E.N.T. clinic and all new up-to-date equipment. The ship used to be the *Iroquois* and has only been commissioned by the Navy since August. There are 13 nurses including the chief nurse who is very sweet. They don't do any bedside nursing, only supervise the 130 corpsmen. . . .

Good night, Darling, Rusty

Following the devastating events of December 7, 1941, the 119 Army and Navy nurses stationed in Hawaii faced exhausting workloads. Although these women sent hastily composed cablegrams and brief letters of assurance to relatives and friends in the States, it was often several weeks before they found the time to write letters that included detailed accounts of the Japanese attack. In her 1942 New Year's letter to her mother, Blanche Kiernan described the events of December 7th and its aftermath.

[Tripler General Hospital, T.H.]
A New Year [1942]

Dearest Mom:

. . . We were visiting . . . at Hickam Field, Saturday night, December 6th. . . . They wanted us to stay all night but we didn't have permission so had to come home. . . . The next morning a bomb landed in the street in front of their quarters and the house was riddled. . . . I don't see how they came out alive. . . . Our old quarters at Hickam had some glass blown out of the windows and a few holes in the roof and it was my day off too. We had just moved into Tripler in time. We had visited Sgt. Bonnie and his wife on Dec. 6th also and had a long chat. He was one of our neighbors and had been married only a month. . . . He was killed the next morning on his way to help out at the hangars. Also Lt. Ritchie who was PX officer and had called on us several times out there was killed as the PX was almost demolished. . . . There have been so many changes that I can't keep up with all of them. Many of the large schools have been turned into hospitals and quite a number of our nurses have been transferred out to staff them. Please take good care of yourself and don't worry about anything. . . .

I love you oodles and gobs, Keep 'Em Flying, Rusty

Blanche Kiernan returned to the United States at the end of the war, and she remained in the Army Nurse Corps until June 1951. In 1948, she married Kenneth W. Kutter. The Kutters had two children and four grandchildren. Blanche Kiernan Kutter died in 1993.

ARMY NURSE MONICA CONTER of Apalachicola, Florida, was stationed at Hickam Field at the time of the Pearl Harbor attack. What follows are letters about the events of December 7th that she wrote to her parents and to Major (later Colonel) Julia O. Flikke, Superintendent of the Army Nurse Corps. The letter to Major Flikke was published in the April 1942 issue of the *American Journal of Nursing*.[2] Monica Conter Benning lives in Ft. Meyers, Florida, and is an active member of White Caps, an organization of military nurses who were stationed in Hawaii on December 7, 1941.

---

2. The letter of Monica Conter is located at the United States Army Medical Department Museum at Fort Sam Houston, Texas. The letter to Major Flikke is reprinted with permission from the *American Journal of Nursing* 42 (April 1942):425–26. Copyright 1942 The American Journal of Nursing Company.

Hickam Field, Honolulu, T.H.

My Dearest Daddy and Mother,                    December 22, 1941

. . . I understand we may write anything that has been published in the local paper concerning the raid and as I know you are curious to know my part in the "show" I will try to give an account.

Sunday A.M.—Dec. 7th I rushed on duty at [censored] late for duty (overslept as we had quite a party the night before at the Pearl Harbor Officers Club). . . . While drinking coffee and tomato juice— I heard some planes real low—one sounded like it might crash on the hospital [censored]. Just as I jumped up from my desk, I heard a *terrible noise*—I said, "A plane crashed"—and ran out on the screened porch, 3rd floor, overlooking Pearl Harbor. The music really started!—What I saw was a lot of black smoke and about [censored] planes so low they looked as if they might be landing in the Harbor. Having lived on the air post I have learned to identify our different types of planes and I know we didn't have anything with big red circles on the side—I turned to some of the patients who know aircraft and the other nurses (Miss Boyd) who *knew* aircraft and said, "My G— it's the Japs!" They laughed and said, "Don't be silly— It's maneuvers." I was beating Sgt. Holliday on the shoulder trying to make him confirm my statement and he laughed too—so I ran downstairs to the Commanding Officer, Capt. Lane, M.C. and asked him if it was the "real McCoy—the Japs." All he could do was shake his head in the affirmative and start making phone calls. Then I got all my patients down stairs to the first floor and we "all" stayed there. I mean the entire medical personnel working.

Really, I never heard so much noise in my life, bombs, some 500 lbs.—machine guns, our anti aircraft—and in the middle of it all some of our [censored] were just coming in from the mainland *without* radio or ammunition or guns. Naturally, a few were "ruined". . . .

The wounded started coming in 10 minutes after the 1st attack. We called Tripler for more ambulances—they wanted to know if we were having "Maneuvers." Imagine! Well, the sight in our hospital I'll never forget. No arms, no legs, intestines hanging out etc. . . . In the meantime, the hangars all around us were burning—and that awful "noise." Then comes the second attack— We all fell face down on the wounded in the halls, O.R., and everywhere and heard the bombers directly over us. We (the nurses and the doctors) had no helmets nor gas masks—and it really was a *"helpless"* feeling. —One of the soldiers who works for my ward saw me and so we *shared* helmets together. In the meantime, the bombs were dropping all around us and when a 500 lb. bomb dropped about [censored] from the [censored], we waited for the plane to come in as it felt like it had hit us—then they were gone. [Censored].

All our electric clocks *stopped* on the dot. The dead were placed in back of the hospital, the walking wounded went in trucks to Tripler, and the seriously injured in the ambulances. We used our place as an "Evacuation Hospital". . . . The mayor sent out 20 cases of whiskey so that helped some—that is, the uninjured who were going around in a daze. —Of course, it was used medicinally too. We worked, and worked, and worked—and when night came on "Blackout" (I'm used to it now). . . . For a week the nurses slept in uniform on the ward in one of the officer's rooms. Then we were moved downstairs to "the X-Ray dark room". . . .

Received another letter from A.D. Glad Daddy wired him. He was quite worried. And I do appreciate people's interest in my welfare. But, tell everyone I wouldn't have missed it for anything. You know, I always loved activity and excitement— For once, I had "enough". . . .

A happy, happy New Year to all—Your loving daughter—Monica

[Hickam Field, Honolulu, T.H.
Late December 1941]

Dear Major Flikke:

. . . We felt that we were the happiest group of nurses anywhere—a new thirty-bed hospital, lovely quarters—just two blocks from the officer's club, and, above all, the grandest chief nurse who enjoys everything as much as we do. . . . Under such pleasant conditions the Japs found us *that* Sunday morning. . . .

We were evacuating the major cases until noon, then for days minor cases due to bomb fragments came in. Water was brought to us in large cans, and the mess had started functioning—in fact, it was the only active one at Hickam that day, and we fed hundreds. Red Cross nurses from Honolulu had come out to help—and officers' wives who were R.N.'s were with us for several days. As you can probably realize, there were parts of that day I can hardly account for.

At sunset, directly in front of the hospital "Old Glory" was still flying even tho she had a huge rip completely across, due to machine gunning and there were several bomb craters a few feet from the mast. Then began our first "blackout" night.

The climate, flowers, scenery, nights, et cetera, are still just wonderful but instead of wearing hibiscus and leis, we are wearing little tin hats and gas masks.

Major Flikke, there are days like December 7 when a nurse can fully appreciate her profession as never before, and deep inside there is a feeling of satisfaction and thankfulness that she was able to do her bit to help "Keep 'em flying."

[Monica Conter], R.N. Hawaii

NURSES STATIONED AT Pearl Harbor were not the only uniformed women to encounter the disruptions of war during 1941. Months before the Japanese attack, nurses assigned to distant postings found themselves caught up in a war that was already raging throughout much of the world. These women wrote descriptive letters about their new assignments. In fact, throughout 1941, the "Letters from Readers" pages of the *American Journal of Nursing* published a variety of letters from nurses for whom wartime conditions had become a tangible reality. What follow are examples of these letters. In the first letter, an ANC nurse describes her part in the famous Louisiana maneuvers of September 1941—the last and the largest prewar maneuvers in the United States.[3]

[Fort Leavenworth, Kansas
November 1941]

[Dear Editor of the *American Journal of Nursing:*]

We left Fort Leavenworth, Kansas, for maneuvers to be held in Arkansas and Louisiana for eight and half weeks. Arriving at Prescott, [Arkansas,] we found the high school building was to be our hospital and our first duty was to make it ready for patients. This meant setting up medical and surgical wards, diet kitchen, laboratory, and operating rooms. We had to make sure that each bed was canopied with mosquito netting and each ward equipped with supplies for immediate use. The diet kitchen was established in the home economics room where all food for the bed patients was prepared. Our sterilizer was a large jacketed kettle heated with gasoline.

The communication quarters, post office, telephone and radio office, medical supply, receiving ward and convalescent wards were in tents set up around the school building, which was the regular field hospital.

Each ward tent had sixteen beds and there was a nurse and well-trained ward boy for every two or three tents. All patients entering the hospital first went to the receiving tent, and they were assigned to a medical or surgical tent, or (the severely wounded) to the main build-

---

3. Reprinted with permission from the *American Journal of Nursing* 42 (January 1942):87. Copyright 1942 The American Journal of Nursing Company.

ing. Patients requiring more than four days' hospitalization were sent to Camp [Joseph T.] Robinson, [Arkansas].

In Louisiana, we experienced "real" maneuvers. As we were subject to air raids, we learned to prepare for the safety and comforts of our patients during the raid. All our medical supplies, food, gasoline and oil, and mail were delivered at night under blackout conditions. The bad weather was quite a feature, one storm almost reaching cyclonic proportions. The rain and mud were a problem, and mosquitoes made it very uncomfortable and caused a constant dread of malaria. Snakes in the woods and marshlands caused some acute emergencies. When a soldier was brought in with a snake bite on his arm, a tourniquet was immediately applied. A T-shaped cut was made at the site of the bite, over which a hot empty milk bottle was placed which acted the same as a suction bulb.

I certainly profited by the experience while on maneuvers. I wish all Army nurses could have the same opportunity.

Dorothy Hicks, Army Nurse Corps


In the next letter, an ANC nurse stationed at an Alaskan outpost reports on the buildup of defense facilities in the northernmost territory of the United States.[4]


[Fort Richardson, Alaska,
Late Summer 1941]

[Dear Editor of the *American Journal of Nursing:*]

The first army nurse arrived at Fort Richardson, Alaska, America's "last frontier," before May first. Our group of four was the second to arrive via Dutch Harbor, the Aleutian Islands, and Seward, Alaska, aboard the *USAT* [United States Armed Transport] *St. Mihiel*. Organization was well under way when we arrived and we found the hospital a smooth-running unit for one so young.

We are located approximately five miles from Anchorage, one of the largest Alaska cities. Because of the defense program, Anchorage is a "boom" town, the population having tripled during the last year.

Alaska's natural beauty is unbelievable! In no place in the world can be found such magnificent sunrises and sunsets. We have had a profusion of wild flowers beginning with wild roses in early spring, followed by shooting stars, baby iris, wild larkspur.

---

4. Reprinted with permission from the *American Journal of Nursing* 41 (October 1941):1,202. Copyright 1941 The American Journal of Nursing Company.

We feel very far from our homes, since there is delay in mail service and we are occasionally lonesome for familiar things unobtainable in Alaska, but we are becoming acclimated rapidly. We feel ourselves a well-knit group because, in addition to other factors, we are isolated. One thing is uppermost in our minds—we are in a section well selected for a concentrated defense area and we are conscious of the fact that we, at all times, must be prepared for whatever comes. We know we are here for a job and we know it will be done! We are glad to do our part for defense on America's "last frontier."

Lillian C. Girarde, Army Nurse Corps

Following the military occupation of Iceland by the United States in July 1941, Army nurses were assigned to duty in that country. In the next letter, an unidentified nurse recounts her early days in Iceland.[5]

> [Reykjavik, Iceland
> Autumn 1941]

[Dear Editor of the *American Journal of Nursing:*]

We have been very busy since our arrival in Iceland, setting up our quarters and issuing the nurses equipment. My hut-mate and I would make good quartermasters now—you should see our housekeeping in a Nissen hut. We are beginning to look quite homelike, and never look at a packing box without realizing its possibilities.

We have made dressing tables, shelves, and closets. The morale among nurses is particularly good and they are all busy as can be trying to outdo the nurses in neighboring huts in getting fixed up. Colorful curtains are appearing at the windows and on the clothes closets.

This is a beautiful country, with scenery much like Hawaii. We can see some beautiful snow-capped mountains from our window. Cannot tell you our location, but have a million dollar view.

The nurses are eager to get to work and all pitch in to get things done. We haven't many sick, so they are making supplies. We are using one hut for a nurses' recreation house, and you should see how they have fixed it up. They crowded in the sleeping huts a little so they could use one for this purpose.

Everyone has been so nice to us. All the Icelandic people I have

---

5. Reprinted with permission from the *American Journal of Nursing* 41 (December 1941):1,447–48. Copyright 1941 The American Journal of Nursing Company.

met have been friendly, and all understand English and speak it, so we have had no difficulties in the stores. There are many enticing book shops.

We wear uniforms at all times. Have put away all civilian clothing.

R.N., Army Nurse Corps

One of the first groups of American Red Cross nurses to be sent to England was associated with the ARC-Harvard University Hospital. In June 1941, a British convoy of ships carrying twenty-nine of these Red Cross nurses sailed for England. Two of the ships were torpedoed by German submarines, resulting in the deaths of six ARC nurses, the first women victims of the Battle of the Atlantic.

ARC nurses in England, dressed in blue serge suits and brimmed felt hats bearing the Red Cross insignia, were extended warm welcomes and expressions of gratitude by the British people. Writing from London in September 1941, an American Red Cross nurse provides a succinct description of a tube (subway) bomb shelter used during the Blitz.[6]

[London, England
September 1941]

[Dear Editor of the *American Journal of Nursing:*]

Last night we went to see the shelters—one in a tube and one, opened only a fortnight, of the very latest type.

Tube stations are on several levels and all much deeper than ours. On the middle tier of all levels is the "medical post," much more like a public health unit than a first-aid post except on a night of a bad blitz. It is located at one end of the platform and is all enclosed including room to keep out dirt. One section is a treatment and consulting room, one has bunks so the Red Cross volunteers can sleep (they work all day at a job and may do shelter work as much as five nights a week). They have all had lectures and at least fifty hours in the hospital wards.

There are usually two volunteers, always a graduate nurse, and a physician part time. There is quite an astounding medicine cabinet. Each bottle labeled with name and dose and whether or not only a

---

6. George Korson, *At His Side: The American Red Cross Overseas in World War II* (New York: Coward-McCann, 1945), 306–8. The letter is reprinted with permission from the *American Journal of Nursing* 41 (November 1941):1,325–26. Copyright 1941 The American Journal of Nursing Company.

doctor may administer it. Outside these rooms on the tube platform
are two three-tiered bunks where the sick or infectious patients are
isolated. Then come the lavatories—all new and all with running water. The tubes do not normally have any toilets. They are very clean.
Then come rows of three-tiered bunks, each numbered. Each person
has a ticket with bunk number. Should he be absent for four nights in
succession his bunk is forfeited. At the opposite end of the platform
is the canteen with cake, meat-pies (hot), cocoa, tea, and a bar of
chocolate for everyone every fourth night. They have to pay for this,
of course. The nurse gets a chance to do a lot of teaching—some
group talks, but mostly individual. She does immunizations, too, takes
temperatures and throat cultures of any apparently ailing ones, isolates
them and reports to health visitors for follow up next day at home.
The health possibilities are enormous. Records are kept. Oh, yes, physical exams every two months on all under five years and neonatal exams are done, too. The one we visited has 700 a night when there is
no blitz, by no means the largest. They play games (darts, checkers,
cards, tag, et cetera). Singing is frequent. No child has ever fallen off
the pavement even playing tag. Adults do occasionally, but the electric
rail is at the far side. They seem very happy. Some elderly people come
just for the company. The ventilation is good, no foul odor at all—
much better in all respects than our New York subways. Of course,
everyone agrees that at first they were very bad. I inquired about deliv-
eries in the tube and found there had been only one in the past year
in this station and many nights there were a thousand people in
it. They call ambulances and get them to the hospital. All pregnant women are urged to leave town before they are due but many
don't.

The new shelter is quite different. The big rooms have been rein-
forced in such a way as to make little cubicles. The families can get
together in the little roomettes—six persons in each. There is a piano,
good books, canteen, lavatory, first-aid, and medical service (much the
same except that medical service does immunizations, teaching, et
cetera).

An American R.N. in England

The following letter was written by Marie Adams, a Red Cross
medical social worker stationed in Manila in November 1941. Adams
was part of a Red Cross team that was sent to the Philippines in the
fall of 1941 to provide services for the growing numbers of military
personnel stationed in the western Pacific. When war struck the Philippines, Adams served in Cavite and Corregidor. In the spring of 1942,

she was imprisoned in the infamous Santo Tomas Internment Camp, where she remained until the end of the war. In this single extant letter, written to Eleanor C. Vincent at American Red Cross National Headquarters in Washington, D.C., Adams describes her initial impressions of life in the Philippines:[7]

[Philippine Islands]
[Dear Eleanor Vincent:]                                                November 20, 1941

The Islands are picturesque and interesting—although I believe I should not care to make them a permanent home. . . . [I hope] that my mind will dry out, and that I will get used to cockroaches the size of frogs running though our quarters. . . . [There are] ubiquitous lizards, millions of tiny ants, and heat, moisture, continual stickiness and drip, drip, drip of my external person. . . .

The Filipinos themselves are fun—I think I shall enjoy working with them as patients. But how they have babies! (Not the patients, of course, —just the wives!) I certainly had an education in the Art of Having Babies the Easy Way while at Sternberg for I was in the maternity ward and marvelled at the rapidity and ease of the Filipino manner of child-bearing! Between the moment when the mother called out, "Mom, my baby is beginning to come" and the moment about two seconds later when the nurse announced,—"Well, you have a nice little baby girl, Manuela"—you scarcely had time to take a long breath! Miss Nau [a Red Cross colleague] attributes this ease and speedy action to the native propensity for "Squatting"—I believe it is due to the fact that that is the easiest way out! Surely things are done the easy way out here, whether that way is the most efficient or not, matters little! . . .

[Marie]

In the last example, an American nurse traveling during the spring of 1941 with an international relief mission on the Egyptian ship *Zamzam* writes of what it was like to be in the direct line of enemy fire.

---

7. Excerpts from the November 20, 1941, letter of Marie Adams to Eleanor C. Vincent appear in Raymond D. Jameson, "The History of the American National Red Cross, Volume XIV, *The American Red Cross During World War II in the Pacific Theater*" (Washington, D.C.: American National Red Cross, 1950), pp. 46–47, 57–58.

She also details her experiences as a prisoner aboard the German ship *Dresden.*[8]

Wilmar, Minnesota
[Summer 1941]

[Dear Editor of the *American Journal of Nursing:*]

Doctors and nurses, American, Canadian, British and Greek, were among the two hundred passengers who sailed from New York, March 20, on the Egyptian ship *Zamzam* bound for Alexandria, Egypt, via Capetown, South Africa. . . .

Accommodations were poor on the ship, sanitation was bad, water was scarce, and the crew members were overworked. There was a large amount of medical supplies and ambulances, clearly marked British-American Ambulance Corps, destined for war areas.

Our first stop was in Baltimore. A week later, March 30, we arrived at Trinidad, received sealed orders from the British Admiralty, and began traveling in blackout. April 8 found us in Pernambuco, Brazil, where more passengers and cargo came aboard. The *Zamzam* then put out for Capetown and for eight days we travelled in radio silence, flew no flag, had no identifying marks, and saw no other ship.

On April 14 in mid-afternoon, the ship swung quickly to the west, heading back for South America, but when dawn revealed an empty sea, we headed again for Capetown. About 5:45 A.M., April 17, five days out of Capetown, we were shelled without warning by a German raider. The ship shook and trembled as salvo after salvo came. Shrapnel burst overhead. The noise was tearing, rending. At least ten shells hit their mark, causing destruction and tragedy; ten persons were wounded, three of them seriously. We jumped quickly out of bed, grabbed a few clothes, purses and lifebelts, and hastened down rope ladders into lifeboats as soon as the firing ceased. The *Zamzam* began to list to port immediately.

Passengers maintained complete calm, but the conglomerate crew was panic-stricken. Two boats filled with water, and the survivors were forced into the shark-infested sea. The swastika waved in the breeze, as the *Tamesis* cautiously came closer, and a voice called out for us to come aboard. Within an hour all passengers and crew had been rescued, and those wounded had been operated on in the ship's hospital.

Food, baggage, and other supplies were transferred from the crippled *Zamzam* to the raider. That afternoon at 2:00 P.M. time bombs placed in the holds went off in quick succession and the Egyptian ship and her $3,000,000 cargo went to the bottom. Most of us lost everything—clothing, books, medical supplies, food, household goods—all intended for years on the mission field.

We milled about the raider's hatch for hours. Babies and children whimpered constantly. At sunset we were herded into the hold, three decks down, and locked in. It was hot and stuffy; filth abounded; men, women and children used the same washrooms and lavatory; there were no clothes for those who had been in the sea. That night was horrible beyond description.

The following afternoon supplies, baggage, and passengers were taken aboard the *Dresden,* the German prison ship which was to be our crowded jail for almost five weeks. We slept on dirty mattresses on the floor with lifebelts for pillows, and one blanket apiece as bedding. Water was rationed—two liters per person per day. Plumbing and sanitation were atrocious. We were locked in from sunset to sunrise, with no fresh air and ventilation. Breakfast consisted of a flour and water paste with occasional grains of oatmeal, and black bread and tea. Later, women and children were given jam with the morning meal. Lunch and supper included rice, macaroni, or bean soup with fragments of corned beef or other meat, black bread, and tea. After many days of pleading, milk and some pureed vegetables were provided for the babies.

The men were given sacks and raw cotton and told to make their own mattresses. They had no washing facilities or lavatory. Each man used his one enamel bowl for three purposes: bathing, shaving, and eating. One hundred and eight men occupied a space 50 by 50 feet in the hold, from which one small ladder led to the above deck. As "involuntary guests" of the German navy, we were not mistreated, but there was always strict and severe discipline. For nine days we milled about in circles, getting nowhere. When the raider returned, we asked to be transferred to a neutral ship, or to be taken to a neutral port as soon as possible. Promises were given us, but they were never fulfilled.

Finally the *Dresden* started north on her incredible run—up the South Atlantic, across the equator, through the north Atlantic, west and north of the Azores, and eastward to Spanish territorial waters. We suffered first from the terrific heat, and then from the dreadful cold. We had not enough clothes to keep warm. Colds, bronchitis, and flu developed; then dysentery because of the bad diet. Few escaped some illness. The ship's doctor used up his small supply of medications. But among men, women and children, not one life was lost due to disease or confinement.

Without lights, without flag, and maintaining a vigilant watch from the bridge, the *Dresden* pressed on. One day the crew began building a barricade around the bridge. All passengers were ordered to remain dressed day and night in case of emergency. Through stormy and heavy seas, we ran the blockade to Nazi-occupied France, where on May 20, we Americans were taken ashore at St. Jean de Luz; sixty fellow passengers including doctors and nurses and the crew were taken to Bordeaux for internment.

During a ten-day stay in Biarritz we were carefully examined by German authorities. At last, release was granted and we left occupied territory on a Spanish train for Lisbon. On June 10, we left on the Portuguese liner, a refugee ship carrying seven hundred passengers and docked in New York on June 21. . . .

Sylvia M. Oiness, R.N.

# A Woman's Place
# Is in Uniform

SOME 16.3 million Americans served in the military during World War II. About 350,000 of these individuals were women. Patriotism and the search for adventure were among the most significant reasons that women joined the armed services. Also important were the many recruitment campaigns that emphasized that a woman who joined the military would free a man to fight.

In 1941, building on the experiences of World War I, when the Navy and Marines enlisted "Yeomanettes" and "Marinettes" to act as secretaries, and the Army contracted with women to serve as telephone operators and dietitians, Representatives Edith Nourse Rogers of Massachusetts and Margaret Chase Smith of Maine, as well as First Lady Eleanor Roosevelt, among others, began to work for the establishment of official women's branches in the various military services. The first of these organizations, the Women's Army Auxiliary Corps (WAAC), was created by act of Congress on May 15, 1942. This organization, though, gave women only partial military status.

On July 1, 1943, following considerable debate, Congress abolished the WAAC and created in its stead the Women's Army Corps (WAC). This new organization provided women the same rank, titles, and pay as their male counterparts. In total, 140,000 women served in the WAAC/WAC during the war years. The women's branch of the Navy, the WAVES (Women Accepted for Volunteer Emergency Service), was created on July 30, 1942. By the end of the war, approximately 100,000 women had served in the WAVES. The women's reserve of the United States Coast Guard, the SPARs (from the Coast

Guard motto, *Semper Paratus,* "Always Ready") was established on November 23, 1942, and 13,000 women eventually joined the Coast Guard. The first public announcement of the Marine Corps Women's Reserve occurred on February 13, 1943, and some 23,000 women served in the Marines during the war years. In addition, 76,000 women served in the Army and Navy Nurse Corps.

While the uniforms of the Army, Navy, Marines, and Coast Guard were most easily discernible, a wide variety of other uniforms were proudly worn by women employed in civilian and quasi-military jobs. In September 1942, two small quasi-military organizations, the Women's Auxiliary Ferrying Squadron (WAFS) and the Women's Flying Training Detachment (WFTD), were established. The WAFS and WFTDs merged into a single organization, the Women Airforce Service Pilots (WASP), in August 1943. Just over 1,000 women wore the santiago-blue uniform of the WASP.

Women who joined the Red Cross, worked as air-raid wardens, and were employed as civilians on military posts often wore uniforms. A uniform was designed for the three million women of the Women's Land Army, who came to the rescue of the nation's crops. Uniforms were worn by members of the Cadet Nurse Corps, the American Women's Voluntary Services, and the women who served with the Salvation Army. The 3,500 women who entertained the troops at United Service Organizations (USO) clubs and shows sometimes wore military uniforms without military insignia, as did the approximately one hundred women who were accredited by the War Department as war correspondents. Girl Scouts, Camp Fire Girls, and members of the Junior Red Cross dressed in uniform while performing their duties. Indeed, the opportunities for women to don war-related uniforms were abundant.

WOMEN WHO ENTERED military service expressed great excitement and immense pride in their new status. Recruits wrote enthusiastic letters to family members and friends about life in the military and the camaraderie they shared with other women in uniform. The first selection of letters was written by WAAC Charlotte Morehouse of Fenton, Michigan. Morehouse was a member of the first officer candidate class of 440 women to report to the WAAC Training Center at Fort Des Moines, Iowa, in July 1942. After an intensive four-week training program, "the pioneer 440" received their commissions. On August 29, 1942, the day of the graduation, General George C. Marshall, Army Chief of Staff, sent a telegram to WAAC Director Lieutenant Colonel Oveta Culp Hobby, which stated: "Please act for me in welcoming them in the Army. This is only the beginning of a magnificent war service by the women of America."[1]

Chicago, Illinois

Dear Mom:                                                            [June 1942]

Will do my best to catch you up to date.

Of course for a week I have eaten, dreamed, slept and breathed nothing but Auxiliary—in fact never suspected myself of so much military furor. This thing gets in your blood—it is a chase, a fascinating contrast, and the farther you go, the more it takes hold on you. However, it's also true that the farther you go the more impossible it seems. I'm enjoying all this with very much the feeling that it is in the lap of the Gods, and I decidedly have no expectations of making the grade. . . .

Mom, I did try to get some work done. Tuesday was mostly getting ready to go to Battle Creek. Had my dental appointment shoved up to that p.m. so that my mouth would be in good shape for the exam, and also crowded in an appointment for hair-set. . . . We reported at 8:30 at Station Hospital. With usual army efficiency (??) they were not set up to take care of us, and things didn't get going for half an hour or more. After that we went through rapidly—chest x-ray, usual clinical exam, dental, eye (they gave me 20/20 or perfect vision *without* my glasses), ear, Kahn and urinalysis, etc. We took time out for lunch at Officer's Mess just across the road and went back. By 2 o'clock I was congratulating myself that I was through—only the final checking-over by the Officer of the Day at the desk. By the way they noticed my tiny varicose veins (which no-one ever even sees) but *never saw* the beard! I spent 2 hours the night before plucking, painstakingly and shaved my tummy 'til it was fairly worn thin. Then wore a coating of liquid makeup on chin and it worked. The notation was "mild acne". . . .

Such things one suffers, and then I won't get in, anyway! . . . Tomorrow morning early I report for the final interview. . . . . .

Love, Charlotte

---

1. Mattie E. Treadwell, *United States Army in World War II, Special Studies, the Women's Army Corps* (Washington, D.C.: Department of the Army, 1954), 72.

[Chicago, Illinois]

Dear Mom:                                                June 23, 1942

Know you're anxious to hear the news. . . . On the most important of subjects there's no news. I simply recall that I wrote you at length and confidently that all was over but a short interview on Thurs. morning. Woe! I reckoned without Gov't. red-tape. Arrived at the P.O. Bldg (which is now a very warlike place with sentries to guard and a line of khaki staff-cars and military chauffeurs waiting for half a block on Van Buren St.) to find a line-up of at least 50 women at the door of 1117! The Chicago ones were sent away to come another day. We from out of town were given *another* set of papers to fill out, *sent* over to another building to have them typed and had identification photos made to attach. Mine, by the way, came out good. . . .

After that it was wait and wait for hours. A Brig. Gen. . . . came out and talked to the group. Told us we were a very select lot. . . . Said that any of us who were not chosen for this first Comp. could surely rate officers' training at once by enlisting in September. They do not yet know exactly how many are to be called to Des Moines from the area. . . . If the number stands at 100 or more I think I stand a very good show, but will not know for awhile and perhaps until the last minute. . . . Will be notified by telegraph by Wash. probably just in time to catch the train. So it goes!

. . . Arrived home, battered, weary and hot about 10 o'clock Fri. evening. The interview was a most satisfactory one—a Brig. Gen. (older man) and a very fine gray-haired woman of obvious big capacities. Another woman—psychiatrist—came to the table and talked with each person just a few minutes. Each interview lasted 15 or 20 mins and I felt—so did everyone I talked with—that they really took an adequate measure of each person and will be perfectly fair in their decisions. Of course, we had no way of knowing we rated, but they were charming, easy to talk to. . . .

I love you, darling. Wish there were an excuse to phone.
Love, Charlotte


Charlotte Morehouse began her basic training with "the pioneer 440" at Fort Des Moines on July 20, 1942. The only surviving letter from the basic training period follows.


Fort Des Moines, Iowa

Dearest Mom:                                             August 19, 1942

Know just what a daze I was in last night—forgot most important item in the letter—your *marvelous* box arrived in excellent shape. Just a few of the butter cookies cracked. Were they *good!* Whole barracks joins me in sending thanks and especially asked me to tell you that. The brownies went around once and I had 1½, which is OK with me and better than average. . . . By the way, is our old iron still in commission? If so, please send. It's badly needed and we can't buy them any more.

Forgot also the richest story yet—one of the girls with a cold reached into her locker in the dark for her Vicks and applied liberally—thought it felt kinda thin, turned on her flashlight and discovered it was the *ink* bottle! P.S.—we now call her "Inky" Lee.

Bye again, Darling—C.

After receiving her commission on August 29, Charlotte Morehouse was transferred to Stuttgart Field near Conway, Arkansas, where she served as Commandant of Students at WAAC/WAC Administrative School No. 3. In a long letter written to her mother in early March 1943, she described her great sense of accomplishment in leading a company of WAACs in a recruitment parade on Main Street in Little Rock. This letter also alludes to the WAAC response to the 1943 "slander" campaign when ugly rumors of promiscuous and immoral behavior about WAACs were widely circulated. One rumor even claimed that 250,000 pregnant WAACs had just returned to the United States from North Africa. At the time of the rumor, fewer than 1,000 WAACs were serving overseas, while the total strength of the Corps amounted to approximately 60,000 women.[2]

                                                    [Conway, Arkansas
Darling Mom:                                          March 8, 1943]

Do you recognize the leading figure in the clipping marked "Arkansas Democrat" as your young hopeful? We thought it pretty good for a newspaper picture. Will keep them to perpetuate my moment of glory! I may never have the chance again. It was the fulfillment of a childhood dream I've secretly cherished all my life—to lead troops in a parade—and I never dreamed I would! All day yesterday I kept stealing glances at the picture, and knew one full measure of satisfaction! At the time, while it was actually going on, I was only conscious of the horde of funny, fleeting thoughts that go through your head—"Try to pin this down—this is a big moment" and then hoping my hat wouldn't sail off (it was a cutting wind that made our eyes water) and

---

2. Treadwell, *Women's Army Corps,* 214.

wishing I had time to blow my nose good before we started, and worrying about car tracks and traffic markers in the pavement and how the girls were coming. . . .

The unit had never marched together before—was just put together Friday night from part of the Detachment and part of the student companies—that, of course, made us all the more anxious. Heaven was on our side, I guess. . . . That, my dear, is a picture to make glad the heart of any infantry officer. Notice that every single *heel* is down—which means the cadence is perfect and the hats are all equidistant, the lines all straight, and (from the shadows on the pavement) all armswings identically the same. And, brother, *that* is *marching*. Except for the colors and band, we were the whole parade, and the line of march was short. It was a cold day, not too big a crowd, but ripples of applause all along the way. We went over big. The whole idea was a recruiting campaign and they say they want us again, in bigger numbers next time.

After the parade, trucks took us out to Camp [Joseph T.] Robinson, where the girls were entertained at the Station Complement Mess and we officers in the Officers' Mess—a good meal. The girls afterward went to the Service Club where a U.S.O. Camp Show was being played and afterward dancing until 10:00. We danced at the Officers' Club. Then struck for home—a cold ride, but everyone happy and singing. The reaction of the [Camp Joseph T.] Robinson men was amusing. Laboring under the delusion that all WAACs were (a) a bunch of toughs, or (b) 43 and homely, they came to the party under protest. 250 were *detailed* to go, told to cancel any other engagements, and threatened with a week's K.P. if they didn't show up! Well, we have a pretty smooth-looking lot of girls. When the party was under way, everyone had a wonderful time; some vowed it was the nicest Service dance they'd ever attended. They plied the girls with questions about their schedule, were amazed at their knowledge of things military, and got a great bang (as we've found that officers do) out of talking shop with females. The M.P.'s had to put the crowd out before we could organize the girls and start home! . . .

Yesterday saw my initial performance as Provost Marshal [Chief of Police]—was called into consultation·by the local police force. Two aux's from Russellville had been dating two of their officers. They all registered at the hotel here—separate rooms however. One girl had been drinking too much, got lost in the hotel corridor looking for the toilet, and stumbled into the room occupied by two soldiers from [Camp] Robinson. She fell against the baseboard, cut her head open and had to be taken to hospital for 5 stitches. It was handled very quietly but could have been a nasty scandal. . . .

*Love* always, Charlotte.

This incident prompted the creation of one of the first WAAC military police detachments. Morehouse, as Provost Marshall, was charged with overseeing the work of the WAAC M.P.s. Relying on her college sociology studies, Morehouse worked closely with local police, and incidents that might have contributed to the "slander" campaign were dealt with easily and equably.

At the time of her discharge from the WAC in 1946, Charlotte Morehouse had achieved the rank of major. She returned to the university under the G.I. Bill of Rights, and later served twenty-six years in the State Department. Fifty years after writing these letters, Charlotte Morehouse remarked, "One thing these letters will do is give you an idea of the quality of ordinary enlisted women in the early days of the Corps. Small wonder that they performed as they did, and left such a memorable heritage. It was a privilege to know and serve with them, and it is a proud and abiding memory."[3]

WOMEN IN THE MILITARY took special pride in their work when important dignitaries visited their duty stations. In the next letter, Lieutenant Margaret Combs Kinsey, stationed in Cleveland, Ohio, describes the exhilarating events surrounding the visit of Captain Mildred McAfee, the Director of the WAVES.[4]

                                                        [Cleveland, Ohio]
Hello, Everybody,                                      March 27, 1943

Tonight I am ending what I know to be three of the busiest days in all my life——even counting those days previous to a Miami High–Miami Edison football game. *Busy* isn't the only word to describe the period of time, perhaps *thrilling, exciting, adventurous,* and *wonderful* would be better. O, you know I am exaggerating! . . .

Lt. Comdr. [McAfee] was to arrive here for two days of programs last Thursday night at nine o'clock. Instead she got in around six o'clock that morning and checked in at the Cleveland Hotel. That afternoon she called one of the Ensigns at the procurement office and

3. Charlotte Morehouse to the authors, June 26, 1992.

4. The letter of Margaret Combs Kinsey is located in the Kinsey, Margaret Combs, USNR, Personal Papers, Operational Archives, Naval Historical Center, Washington, D.C.

the ensign refused to believe that it was she. After much persuasion Ensign Landis decided it was. In a few minutes I was on the telephone trying to notify our Hotel Hollenden that Miss McAfee was being transferred to the suite that we had reserved for her on the tenth floor. . . .

From then and on things just happened so fast that I can't tell half that did happen. All I know is that rather frequently I would pinch myself on the arm and say to myself or to a friend near by, "Is this really you, Margaret?" I can't believe that it all happened to me. I was Miss McAfee's guest in her room for two hours. We just about reconstructed the entire Navy for women. Of course, none of you who know me can imagine how calm and uninterested I was!!!!! My enthusiasm and spirit rose so high I could hardly keep my feet on the floor.

The next morning I just put an extra girl in the office to keep up with me and with where I was to be in order to keep up with Miss McAfee. Perhaps I should explain that she was really in the city in behalf of the WAVES and SPARS campaign; therefore she was under the Procurement and Recruiting Office rather than under our Field Branch. I was simply the honor officer because I was the senior officer at the Branch. Friday at noon the Women's City Club honored her with a magnificent luncheon with about three hundred guests. My Admiral McMillan was among the honored guests. Well, that gave me another thrill. In the Navy no one ever goes anywhere with the Admiral but just appears there afterward. To my surprise, he walked out to my desk about 11:30 and asked where the luncheon was to be and what time I was going. I thought nothing of it until 1150 he walked out with his coat on and said, "Miss Combs, are you ready?" In the Navy or not, when your superior asks such you reply, "Yes, sir." Then my knees began to shake for I hoped I wouldn't make any mistakes getting in and off the elevator, going out doors wrong, walking on the right or left side, and a million other things, for I had never walked three steps with him before. I just silently said, "Now, knees, you've held me up before. You can't afford to let the Navy down." It was a marvelous experience, for he just shoved doors open for me. Put me on and off the elevators first and I felt like a debutante making her first appearance in public. I learned that he was just a great person who wanted to treat me as if I were a lady and not a uniform. Needless to say, the speech by Miss McAfee was a superb one—if you heard her introduce Madame Chiang, you know what I mean. Her wit is superb.

Friday night all of us WAVE officers entertained her in the Ballroom of the Hollenden Hotel with a banquet. All the enlisted girls came up too and sang her songs that they had written for her. Then she made another speech. At eight o'clock the city had planned a parade. All I knew was that I was to ride in an official car. There I was sitting in the back of some Cadillac or something with a city commissioner and Lt. Gray of the Marines (Woman). When the parade started we thought we would be chatting and visiting all the way down the avenue. To our utter dismay, people were so thick every where and we were so busy saluting and recognizing applause that all we could do was smile. Just think of it! 75,000 people applauding us. Bands and Bands!!! Made me think of New Year's Day in the stadium in Miami. Then we were all deposited in the park on the SQUARE outside, believe me, and a gorgeous night, the warmest since I came here. There I was sitting between commanders and admirals in the Navy and Coast Guard. But I almost "busted off all six buttons of my top coat" when my 118 enlisted girls came marching in the Square in beautiful formation and making figures. Gee, they were wonderful. Cleveland gave them a hand of applause that will ring forever in my ears. I marched to the front of the platform and received the salute. Chills just ran up and down my spine. But I didn't get tears in my eyes until I stepped back in place and a high school band in red and white came marching up in front and played marches. . . .

At ten o'clock this morning with numerous escorts, I met Miss McAfee at the Navy building entrance and had her escorted to our Admiral to make her official call. After that with an escorting and official party we inspected for two hours the building and the girls at work. . . . At 3:45 Miss McAfee slipped away for awhile, and I ran back to the Hotel to make plans for a private dinner here in the dining room for five of us officers and her. To our utter dismay, we were her guests. But the visit couldn't just end so calmly. . . . At five minutes before the train pulled out, she took her ticket out of her pocket book and it was for the Penn. train. She hadn't looked at it before. Because she came in at that station she thought she left that way. . . . Well, we made a mad break. Ensign Landis ordered the U.S. Mail Wagon out of the street. I ordered the taxi driver to move on, and the Navy limousine pulled out with three minutes to get there. . . . Just as we drove up, the train pulled out. One person got out and ran in to have it held at 55th street,—just fifty blocks out. We found all the red street lights in the city and they all dared us to move faster. We got there and started up the steps. The station master yelled, "O, there you are." Yes, *THERE*. He started yelling at the train and we started yelling and signalling. Everyone on the last two coaches yelled at us and waved, but we didn't stop the train. . . . [McAfee eventually caught a later train.] Where is Miss McAfee now? I hope someone in the Navy knows!

I forgot to say that I took Miss McAfee around the barracks—fourth floor of the Hotel Hollenden at 5:30. There are a few vacant rooms on this floor, so I thought I would show her a choice one or two. When I pushed open the door I heard a man's voice, apologized, and withdrew. I thought I recognized the voice of one of the assistant managers. When I came out of the dining room after supper, the manager walked up to me with a very serious expression—"Miss Combs, Mr. Skelley, our assistant manager wishes to see you." I replied that I was sorry but we had to catch a train. I immediately bumped into the man. I learned that I not only locked the man in the room, but that I put all the lights out. When you lock these room doors from the outside all the lights go out. He was in there shaving. When I told Miss McAfee that one, she about went into hysterics. O, yes, we initiated her. . . .

Well, this has been just another of my bombastic moments. Excuse me for taking so much of your valuable time. But some of you will sigh and be glad that you didn't *have to* listen to me speak at the banquet. You can always be grateful to Hitler for relieving you of me and my "bombastic" moments. . . .

"Join the Navy and see the world" has been the slogan of the men. But ours is "Join the WAVES and live." . . .

Love, [Margaret]

The question of what to call the women's organizations of the various military services was the subject of much discussion. In a memorandum of November 14, 1942, Captain Dorothy C. Stratton, the first director of the SPARs, explained to Vice Admiral Russell R. Waesche, Commandant of the Coast Guard, why she preferred the term SPAR for women who served in the Coast Guard.[5]

CG-WR-057
14 November, 1942

<u>MEMORANDUM FOR THE COMMANDANT</u>

1. Perhaps we have a solution to the insistent problem of a name for the Women's Reserve of the Coast Guard. The motto of the Coast Guard is "Semper Paratus—Always Ready." The initials of this motto are, of course, SPAR. Why not call the members of the Women's Reserve SPARS? In the press releases, the press would emphasize and publicize the motto of the Coast Guard. Moreover, there is no problem of inventing titles to fill out the word. As I understand it, a spar

---

5. The Dorothy C. Stratton memorandum is located in D.C. Stratton—Memo SPARs—MVP—629, United States Coast Guard Museum, New London, Connecticut.

is often a supporting beam and that is what we hope each member of the Women's Reserve will be.

2. It seems to me that there is no possibility of avoiding some catch name in the press. If we do not create a name, we shall be called WARCOGS or something worse. I like SPARS because it has meaning.

3. Will you please give me your reaction.

Dorothy C. Stratton

Following the formation of the SPARs, as well as the WAAC/ WAC, WAVES, and Women Marines, large numbers of women began to volunteer for military service. In each of the women's military organizations, the minimum-age requirement was twenty. The fact that men of eighteen could join the military or be drafted did not escape the attention of young women under twenty. A letter written by a sixteen-year-old high school student who hoped to join the SPARs was published in the column "The Mail Buoy" of the October 1943 issue of the *U.S. Coast Guard Magazine.* It perceptively expressed the concerns and disappointment of the young women who did not meet the twenty-year-old age requirement.[6]

Brooklyn, New York
[Dear Editor:]                                    [September 1943]

Why aren't girls of seventeen and eighteen permitted to join the SPARS, while boys of seventeen are allowed in the Coast Guard? Is it fair to make girls between seventeen and twenty sit at home and pray for their twenty-first birthday to hurry, while the boys of the same age, (and mentally, if not less) are fighting the enemy in rough and dangerous waters? No! It is absolutely wrong. There is a saying, "All is fair in love and war," but, honestly, is it fair for us to twiddle our thumbs, while boys are getting their arms and legs shot off?

I am a girl nearing my seventeenth birthday. I expect to finish high school in a few months. What am I going to do after I graduate? Buy war bonds and stamps? We all know that is not enough. I am a normal, healthy, American girl, willing to do more than buy stamps and bonds. Of course, I can join the Red Cross, or the American Women's Voluntary Services, but why not leave that for the older women. I want to join the SPARS, knowing I am taking the place of a Coast

Guardsman, and relieving him for actual fighting duty. I want to be an aid to my country, directly, not indirectly.

If my brothers and friends are old enough to join the Coast Guard and Navy at seventeen and up, I certainly should be permitted to join very soon, on my seventeenth birthday. So, please, who ever is in charge of the Spars, please lower the enlistment age to seventeen and make my seventeenth birthday the happiest.

I know I am speaking for millions of girls who feel the same way about this subject.

Shirley Band

The establishment of the WAAC/WAC, WAVES, SPARs, and Women Marines generated considerable debate. Both inside and outside the military, significant objections to women's performing "men's jobs" were voiced. Moreover, neither young men in the military nor their loved ones were necessarily enthusiastic about women's joining the armed services in order to release men for combat. During the latter months of 1943, these types of concerns were expressed in a series of letters written by Coast Guardsmen who were disdainful of women in the military. The letters were printed in the *U.S. Coast Guard Magazine* and prompted strong replies from SPARs, which were also published. Eventually, Radioman First Class Paul Alexander, the author of one of the original letters of complaint, wrote a retraction in which he described how his viewpoint had been widened by the letters from the SPARs. What follows is one SPAR's response to Paul Alexander.[7]

St. Louis, Missouri

Mr. Paul Alexander:                                    [December 1943]

My dear Mr. Alexander—I have just completed reading your letter in the December issue, to the editor about the SPARS.

I, speaking for all of us, am truly sorry you feel the way you do. None of us joined the U.S. Coast Guard because we thought it was glamorous, or because we had a mercenary idea in mind. Do you imagine that it is glamorous to wear the same uniform day in and day out? Don't you think that we would make quite a lot more money in

7. "The Mail Buoy," *U.S. Coast Guard Magazine*, January 1944, p. 7.

civilian life, right now? (Or didn't you know that we receive the same pay as you boys?)

In a way, you appear to be a rather pathetic creature who like a child is whining because he can't have his cake and eat it, too.

Personally, I joined the Coast Guard because a Coast Guardsman (a true Coast Guardsman) rescued my father during the last war. Somewhere today that same man is fighting for all of us. At the time of my enlistment I remember his words in a letter: "God bless the women of America, and may God especially bless you. You've chosen a top-notch outfit—be as proud of it as I am of you."

Until reading your article today no one has ever shaken my faith in the boys of the Coast Guard.

In closing I say: "God Bless you in your work and may He bring you safely home again with more laudable thoughts of the Coast Guard SPARS."

Sincerely, Mary B. Coolen, Y3c [Yeoman Third Class]


The privilege of serving in the SPARs as an officer inspired Carolyn B. Lundberg to write a stirring letter to her Navy husband, who was serving in the Pacific. She wrote this letter shortly after her graduation from officer's candidate school at the United States Coast Guard Academy in New London, Connecticut.[8]


[New London, Connecticut]
Dear Don,                                          17 May 1944

Pledges inspired by graduation day and preceding indoctrination.

1. May my rank be a constant reminder of my responsibility.

2. May my work and behavior be worthy of the honor I've won by being commissioned an officer in the U.S. Coast Guard Reserve.

3. May I have the moral courage to put the service first, even before you, since this and similar services will return you to me the sooner—and reunite millions of people.

Bon Voyage, Carolyn


The Marines were the last of the service branches to admit women. Indeed, Marine antipathy toward women in the military was strong,

---

8. The letter of Carolyn Lundberg is located in SPAR, ADY – 9, 90.38, United States Coast Guard Museum, New London, Connecticut.

and, according to the standard work on the history of the Marines, "there was considerable unhappiness about making the Corps anything but a club for white men."[9] Marines tell the story of how, at a dinner party in the late autumn of 1942, given by World War II Commandant of the Marines General Thomas Holcomb at his official residence, the discussion at the dinner table turned to the question of women's becoming members of the Marine Corps. The general was able to inform a female guest that the decision to allow women in the Marines had just been taken. At that precise moment, according to General Holcomb, a portrait of General Archibald Henderson, Commandant of the Marine Corps from 1820 to 1859, fell off the wall and landed on a Japanese tea set! General Holcomb later commented that General Henderson was evidently "irritated" by "an announcement which he doubtless resented."[10]

Yet following the admittance of women into the Marines early in 1943, General Holcomb remarked, "There's hardly any work at our Marine stations that women can't do as well as men. They do some work far better than men. . . . What is more, they're real *Marines*. They don't have a nickname, and they don't need one. They get their basic training in a Marine atmosphere, at a Marine Post. They inherit the traditions of the Marines. They are Marines."[11]

In recognition of the important work of the women in the Corps, Major Ruth Cheney Streeter, Director of the Women Marines, sent an open letter to her "girls" at boot camp in October 1943 and stated, "It is not easy to 'free a Marine to fight.' It takes courage—the courage to embark on a new and alien way of life. . . . Your spirit is a source of constant inspiration to all who work with you. Your performance is a promise not only of a victory in the grim struggle in which we are engaged, but . . . of a better world than we have ever known before."[12]

The following letter, written by Rosie Katz of Bloomfield, New Jersey, depicts the special esprit of Women Marines. Katz was a radio-

---

9. Allan R. Millett, *Semper Fidelis: The History of the United States Marine Corps* (New York: Macmillan Publishing Co., 1980), 374.

10. General T. Holcomb to Stuart E. Jones, 3 April 1958. Shoup Papers, Box 1A222, Marine Corps Historical Center, Washington, D.C.

11. Pat Meid, "Marine Corps Women's Reserve in World War II" (Washington, D.C.: Historical Branch, U.S. Marine Corps, 1968), p. 64.

12. The Streeter letter is quoted in Peter A. Soderbergh, *Women Marines: The World War II Era* (Westport, Conn.: Praeger, 1992), 53.

man, and, at the end of the war, she was transferred to Ewa Air Base in Oahu, Hawaii.

[Cherry Point, North Carolina]
Dear Folks:                                         November 10, 1943

As I write to you today on the 168th anniversary of the Marine Corps my thoughts wander way out to the South Pacific knowing that the boys out there are fighting with all their might so that on the next anniversary of our corps they may be home celebrating it in the splendor it rightly deserves.

You might think I've gone code nutty [radio school was filled with rumors of people driven around the bend because they constantly heard Morse code] or something, but don't worry I haven't. It's just that I'm reading an excellent book about my buddies in the South Pacific [Walter L. J. Baylor, *Last Man Off Wake Island,* 1943], and it sort of made me realize how fortunate I was to be accepted in such a wonderful Corps. I'll send the book home just as soon as I finish it, probably today. I found that the writer, Lt. Col. Walter Baylor, whose experience on Wake, Midway and Guadalcanal are related in the book is now a Colonel and stationed right here at Cherry Point. I'd like to get him to autograph it, so I'm going to ask one of our Lieuts. the best way to go around to seeing him without violating military procedure. Reading the book, has changed me so, really, it has. I thought that when I joined the Marines I knew why I joined, but not until I read this book, did I find the real answer.

I'm glad I'm a Marine, but, I only wish I were a male Marine instead of a female Marine. But one thing I'm thankful for is that I'm in communications. This officer was in communications, setting up radio communications between planes and ground at the various islands. There's a paragraph in which he says that a radioman "has exciting and important duties that require steady nerves and quick thinking while under fire. It's full time work, and no job for weaklings. If a man wants action, let him join the Marines and get into communications. He won't be disappointed." Well I read that paragraph about 3 or 4 times. Just think, if, God forbid, we should ever be attacked it'll be my job to contact the planes and tell them just where the enemy is coming from, etc. Boy, I'm happy to be a radio operator in the Marine Corps.

Let's not be sentimental anymore. I hope you are all well. I'm just fine and happy. Tomorrow nite I go to that steak dinner at an Army base. Oh, boy! . . .

Going to finish my good book now. So I'll write you tomorrow. Take care of yourselves, and be good. Loads of Love and Kisses.

Your loving daughter and sister, Ro

Shirley Magowan of Springfield, Massachusetts, one of the first Marine recruits, joined the Women Marines in early 1943. Because the Marine Corps had not yet readied its own training schools, she took her basic training at the Hunter College Naval Training Station in the Bronx, New York, with women who were WAVES. In July 1943, the Hunter College training center for Women Marines was transferred to Camp Lejeune in New River, North Carolina. Following basic training, Magowan was sent to Storekeeper School in Milledgeville, Georgia, and then to Quartermaster School at the Marine Corps Air Station in Edenton, North Carolina, where she remained until her discharge at the end of the war. Shortly after the war, she married Bill Crowell, an enlisted Marine she had met while at Edenton. Shirley Magowan Crowell lives in Tucson, Arizona, and is an active member of the Tucson Women Marines Association. As her letters indicate, she, too, was thrilled with her new military life. The following letters, which describe her first few months in the Marines, were written to her sister, Gail.

                                              [Bronx, New York]
Dear Gail,                                      May 19, 1943
. . . The scuttlebutt has it that future Marines go to New River, N.C. This place is so Navy you think you'd joined the Coast Guards or the Waves. Our immediate superiors are girls who've been here five weeks and our company commander is an Ensign (Navy Wave). . . .

Two days have elapsed. I had my first Guard duty at 4:00 AM Thursday morning. We were rushed all day Wed., given no time off, not even our quiet period at night. Up at 3:30 and made our bunks in the dark. Got the 2nd deck up at 5:30. Hit the deck! Did I feel like a meanie. We had our shots—three of them—Wednesday and were all rather wish washy. No one was sick. Wanda had a few chills. Cleaned room by 6:15, headed for Mess. We have to walk 20 minutes to and from mess, classes, drill and gym. That's a total of 40 minutes marching, and we trot back and forth all day long. We had a map reading class, went to the gym, had inspection (all got bawled out), drilled (tough men Marine Sgts) more class, home "M", worked hard scrubbing, mess and the evening off (7:00 to 9:00) they had promised us developed into tests which lasted two hours. We were all dead tired. I was going to sick bay at 7:30 (you can only get sick twice a day 8:30 AM and 7:30 PM). I got by the physical. It was a cinch. They just glanced at us and copied material from Boston. In some instances only, of course. I was sick Thursday night but am improved today. I could actually swallow some food and did make up for yesterday! Wow! I've eaten everything in sight, yes, even tomatoes and coffee with condensed milk.

Today has been good. Feeling tops makes the grind even pleasant. I like the life, even if it is exhausting. We've been plodding around in the rain all day. It's their pet habit to leave us standing out in the elements and today we got caught without our rain gear and we were drenched to the skin. Couldn't even change when we got back.

Got fittings for uniform. Don't faint now, they're sticking me in a 11½, because I'm a shorty. 'Cause I'm so "little" they didn't have anything to fit me, it had to be ordered. . . .

With love from your little sister, Shirl.


[Aboard troop train in Virginia]

Dear Gail,                                                    June 13, 1943

So many things have happened so fast the past few days my head is whirling. Right now I'm sitting on a train passing through Virginia. We are starved. It's after nine and we haven't eaten since four yesterday afternoon. We have eaten now, and boy, did it taste good! Bacon, eggs, potato, coffee, and cereal. Everyone was sick before breakfast and I felt swell. Now I'm burping all over the place.

There are [censored] WAVES and [censored] Marines heading for the school. We are travelling on a troop train. That doesn't sound so glamorous but wait till I tell you about it. Your little sister is always on the tail end of everything, because she's short. This time it really paid off. We are going by Pullman and a girl I just met and I got a roomette. It accommodates three, but there were only two mattresses. We have our own private john and bowls. There is a long couch we can stretch out on. Eight could sit here comfortably. I slept in the upper and it was more comfortable than my bunk. I would have been hot but there was a fan that we had on all night and I had my cover on by morning. . . .

We will get stiff exams when we leave this [storekeeper] school and approximately 10 of the hundred will get corporals ratings, a few will get P.F.C. stripes and the rest will be P.F.D.'s (private for the duration). All joking aside, they say it is plenty tough, and that we're lucky to be going. The only drawback to our good intentions will be the heat. My arms, back and face are all wet, and my skirt and stockings are sticking. We just had to sign slips as to our typing experience. It asked about figures and I put down that I typed them very slowly. I don't want to have to do that all day long. . . .

I'll stop my foolish prattle for by now I'm probably beginning to bore you. Be sure to write and tell me all your plans.

Bye for now, and keep smiling.

Love, Shirl.


                                              [Milledgeville, Georgia]
Dear Gail,                                         June 17, 1943

. . . . Studies terrific. 1 hr typing a day and 5 hrs of our subject, General Storekeeping. Drill or gym daily, 1 daily hr. liberty, or free time and 1 hr compulsory study period. We are cramming, but it is terribly hard. I've learned more in 2 days than I had to at Hunter in 3 weeks. . . . We're all doubtful of passing. . . . But, if we manage to get through the rewards are pretty nice. I'm trying to wake up mornings to study, it's the only time I can concentrate. . . .

I have as much information crammed into my head for this Saturday's exam as I ever did for a final in a H.S. subject. No fooling, we're just swimming in figures, titles, accounts, funds, appropriations, allotments, etc. We have to be able to identify 8 or 10 of the above (each) by either number or names. There are 25 stock forms to know, numbers and names and a million other definitions, organization, structures, etc. Have said enough?

They are going to make an accurate typist of me. My rate, 55 per min, for 5 minutes has gone up 10 words so soon with same 2 errors. We are learning to type figures—something I never did properly and I'm having trouble. It will come in a while—I hope. . . .

Love from your family, Shirl.


When nurses entered the Army and Navy Nurse Corps, they had little knowledge about military life or the combat conditions many of them would soon encounter. In July 1943, the Army authorized special four-week military orientation programs for newly commissioned nurses. From its founding until the end of the war, over 27,000 nurses were graduated from these programs. Helen K. McKee of San Antonio, Texas, was one of the first nurses to undergo this special training. Her letters to her parents, written from Camp Forrest, Tennessee, prior to her assignment on the Italian front, describe this orientation program. (For other letters by McKee, see pages 130–136.)

Camp Forrest, Tennessee
Dear Mother and Dad:          May 6, 1943

Who dreams up all these little fun and games activities we have engaged in lately? The latest was a ten mile forced march with all manner of diversions en route. . . .

But the real fun comes in marching with the gas mask on. You can get an adequate amount of air, but none to spare, especially when you are force marching. That was only part of the entertainment provided us. Someone came up with the idea of giving the air corps a little practice in bombing, and us a little practice in ducking. . . . You run for cover, throw your gas mask back and fall, your elbows catch your body, and your face remains off the ground. This is to prevent concussion from exploding bombs. . . . I know definitely I was killed at least three times that day. . . . When we finally reached our destination, Cumberland Springs, we climbed down a hill, took off our shoes and sox and waded in the cool spring water. My, but it felt good. We then had our rations which consisted of two sandwiches, coke or coffee and cake. Hardly an adequate meal for a hard working soldier.

Mother and Dad, I think you are to be congratulated on producing such a sturdy offspring. . . .

Deepest love to all, Helen


Camp Forrest Tennessee
Dearest Ones:          July 18, 1943

After doing a ten mile forced march with full field pack some time ago, we thought we had triumphantly met the Army's ultimate challenge. But the final test of our physical stamina was to be encountered on the infiltration course. . . . It was about the most thrilling, completely rugged adventure of my life.

First, there was the matter of suitable attire. I sought out the smallest soldier in the 300th, and borrowed his fatigues. At that my arms and legs were lost, so I borrowed a pair of leggings to hold up the pants legs. I soon mastered the trick of wrapping them around my legs. Next came the assembling of our field equipment: canteen, gas mask, pistol belt and steel helmet. What a motley assortment we were as we climbed into trucks to be taken to the infiltration course, which was an area one hundred yards in length with a trench at both ends. The trenches were four feet deep and half filled with water and mud. At one end, overlooking the course, were two towers from which machine gun fire and explosions were directed, and from which blared forth directions and warnings. . . .

We were allowed a few minutes of "at rest" following our unloading from the trucks. This enabled us to adjust to the assault on our ears of the rat-a-tat-tat of the machine gun fire and the boom of the dynamite explosions. Then an infantry officer gave the orders, "Fall in." We fell in. "Right Dress." We dressed right. "Front." We fronted. "At ease." We were at ease, and he went over the plan and the procedures of the course. He demonstrated the best infantryman crawl, and then explained that machine gun bullets would be aimed two feet above our heads. This bit of information seemed to make the subsequent instructions superfluous; but nevertheless, he continued, "Keep your heads down, your hands, arms and feet down." Then he thundered, "And keep your fannies down." We adjusted our helmets, made sure our pistol belts, canteens, etc., were correctly positioned. "Attention," came the next command, "left face; forward march; route step; halt." We were standing at the edge of the starting trench. Here he instructed us to crawl down into the ditch. It was about waist deep in slush. As we parted he warned us to take our time, adding that the infantryman's average time for the course was twelve minutes, but that he expected no better than thirty minutes from us. We were then on our own.

The signal to start firing followed, and my troubles began. As instructed I braced my body on the front of the trench, but my legs were too short to reach it, and that is the only way you can go over and still hug the ground as you emerge. So I waded down to a point in the trench which appeared a little narrower, and tried once more. This time I succeeded. Once out of the trench I found that the mud was no longer my enemy, for it was easier to slide in the slush and slither along the ground than to pull one's body over dry terrain.

In the meantime the rat-a-tat-tat became a part of you and you knew the boom would follow shortly, and you sensed that all was well and you were on your way. I think each of us was filled with a grim determination to finish and in good time. After about fifty years of crawling on my tummy, I found the first strand of barbed wire which was stretched across the course at the half-way point. I lay still a few seconds rehearsing the procedure; then deliberately rolled over on the side opposite my canteen, etc., onto my back, reached behind me, drew the first strand of wire forward, moved my body through, letting the strand go as I reached for the next strand. Just at that moment a blast went off nearby and I was caught facing upward. A shower of mud came down on my face. Since I was in barbwire entanglement there was no choice except to let go and continue. I clearly remember counting nine such strands of wire that had to be negotiated.

With that obstacle behind me I again rolled over and set out for the remaining fifty yards, slithering through the mud. We were beginning to feel muscle strain and becoming aware of scratches and bruises around elbows, knees and hips. I heard a booming voice from the tower, "Sergeant, move to the right. You are headed toward a mine." The message came through a second time before I realized I was the endangered species, for the attire I had borrowed for this occasion bore a sergeant's stripes.

At last my outstretched hand caught the bank of the trench. I maneuvered around to a side position and rolled over the bank into the trench. The muddy water felt almost refreshing after expending so much energy in the intense mid-afternoon heat.

Ninety percent of the girls, including myself, had completed the course in fifteen minutes, and we had been allowed thirty. We sat there neck deep in water until the "all clear" signal was given.

The gun fire had ceased and now new sounds greeted our ears. An army band had been trucked in while we were so intently involved in covering the course, and as we emerged from trench, covered with mud, the band struck up with "O, You Beautiful Doll." . . .

My deepest love to all of you, Helen

The thousands of women who were assigned to overseas duty with the American Red Cross were also required to undergo Stateside orientation and training sessions. Betty B. John joined the American Red Cross in August 1942 and received her training in Washington, D.C. Later that year, she sailed to England, where she served as a Red Cross club director until the spring of 1944. She spent the remaining months of the war as a war correspondent on the European continent. The following letters were written to her husband, Colonel Henry J. John, a physician in the Army Medical Corps, who was stationed Stateside throughout the war. (For other letters by John, see pages 146–150.)

Washington, D.C.

Henry Dearest:                                                    August 9, 1942

. . . I guess its only natural to be so down—having suddenly become completely uprooted after the stability of nearly 15 years. But that is what war does to some people. Should I have waited to do this until after a Fall spent with you? I couldn't just live in Atlanta for good— and if you were moved on—I'd hate having to be constantly on your hands—or if you were sent overseas—I'd hate to have been stranded in Atlanta—or anywhere. I'd much rather be active—even to the point of annihilation now—for I can't see much of a future for any of us if things go on in this manner. . . .

Love, Betty

                                                    Washington, D.C.
Henry Dearest. —                                    August 11, 1942

Even tho' it's nearly midnight and I've just finished studying my lessons (and I mean lessons—this is no picnic) and am miserable with a
soupy cold—I just had to write. . . .

    Classes are all general and have training and basic training and
seem to have little to do with what I'm to do eventually—except that
out in the field—if some one of staff falls ill or is recalled, one has to
be ready to pinch-hit in almost every capacity. There's a cross section
of the country's social and financial strata in our 2 week's class—63—
the one ahead of us had 85— Some 1500 have been trained since
January for these top jobs here and abroad—and there's a lot to it.
They invest a good deal in our training and no monkey business as a
result— The night study is the hardest—and I was in class until after
six today. . . .

    Next— After these two weeks—I will have an extra three day ses-
sion on my own specific unit and work—here—and two weeks in some
post in this country—*but* that I could be pulled out of class to-morrow
or any time before these two weeks are up and be sent with a unit
about to leave. They are that desperate for the type of recreational
training I've had— So each letter may be my last—or you may be
hearing from me for weeks—even months if I become the "forgotten
man"—(which isn't likely). . . .

    How is the Red Cross unit at your hospital? Do you know any of
them down there at the Naval Base? One of my instructors was at the
Base for a while as Field Director. It would be wonderful if I could
be sent down there. Such is life—and it never will be normal again
until the war is over and we're at home together again.

All my love, Betty


    Entrance into the military provided many new job opportunities
for women. Although a large percentage of women in the military performed administrative and clerical work, many other employment possibilities existed. WACs worked as carpenters, cartographers, censors,
electrical specialists, photographers, postal directory workers, weather
forecasters, and X-ray technicians. SPARs held forty-two different enlisted rates, ranging from Boatswain's Mates, Parachute Riggers, and
Control Tower Operators, to Pharmacists. WAVES and Women Marines worked as draftsmen, painters, recruiters, truck drivers, and
welders. The following letters describe some of the less traditional jobs
held by servicewomen during World War II.

    Lorraine "Bub" Turnbull of Rockdale, Wisconsin, joined the

# Part I

## Recruitment, Training, and Stateside Service

When the U.S. entered World War II, the only opportunities for women to serve in the military were as Army and Navy nurses. This illustration *(left)* appeared on the cover of a January 1941 American Red Cross pamphlet encouraging women to join the Army and Navy Nurse Corps as part of the nation's preparedness effort. *(Courtesy of Monica Conter Benning)*

Lt. Monica Conter *(right)*, one of two Army nurses on duty at the Station Hospital, Hickam Field, Hawaii, when the Japanese attacked Pearl Harbor on December 7, 1941. *(Courtesy of Monica Conter Benning)*

Despite significant objections to women's performing "men's jobs," the Army, Navy, Marines, and Coast Guard had established women's divisions by early 1943. *(Authors' collection)*

Two examples of recruitment posters.
*(Authors' collection and U.S. Naval Historical Center)*

Although many servicemen initially expressed misgivings about women's serving in the military, each of the service branches featured women in uniform on the covers of its magazines. A photograph of Ensign Janet Mary Murray, one of 11,086 women to serve in the Navy Nurse Corps during World War II, appeared on the cover of the August 1942 issue of *Our Navy*. (*Courtesy of Robert Wooddall*)

Thirty-nine African-American women were among the "pioneer 440" who reported to Fort Des Moines, Iowa, for officer candidate training for the Women's Army Auxiliary Corps in July 1942. Pictured here are Lts. Harriet West and Irma Cayton, two members of the First Company, First Regiment of the WAAC, going over their recruiting reports at WAAC Headquarters in Washington, D.C., in the fall of 1942. Throughout the war, African-American women served in the WAAC/WAC on a segregated basis. (*National Archives*)

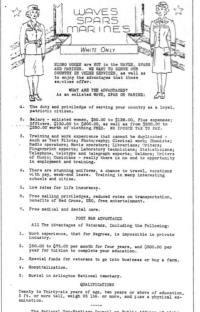

Black women were not accepted in the WAVES and the SPARs until November 1944, a change that occurred only after substantial effort and protest by progressive organizations. Both WAVES and SPARs were fully integrated in their service branches. Neither the Women Marines nor the Women Airforce Service Pilots accepted African Americans. *(Nannie H. Burroughs Papers, Library of Congress)*

Women Marine recruits take the oath of office. *(Ruth Streeter Collection, Marine Corps Historical Center)*

Five Navy nurses take the oath of office. Phyllis Mae Dailey, the Navy's first African-American nurse, is second from the right. March 8, 1945. *(National Archives)*

---

One of the first groups of WAC recruits from Puerto Rico aboard an Army transport plane on the way to the United States for basic training. October 1944. *(Women's Army Corps Museum)*

Representative Margaret Chase Smith of Maine, the only woman member of the House Naval Affairs Committee, was a strong proponent of women in the military. During a May 1944 inspection of the Marine Corps training center at Camp Lejeune, North Carolina, she was instructed in the use of a carbine. Onlookers include Colonel Ruth Cheney Streeter, left, Director of the Marine Corps Women's Reserve, and Major Katherine A. Towle, Assistant for the Women's Reserve at Camp Lejeune. *(Margaret Chase Smith Library)*

Women Marines, in training at Camp Lejeune, North Carolina, stand at parade rest. 1943. *(National Archives)*

Two African-American SPARs pause on the ladder of the dry-land ship, *U.S.S. Neversail,* during boot training at the U.S. Coast Guard Training Station, Brooklyn, New York. *(National Archives)*

WAVES at chow while in training at Norman, Oklahoma. February 1943. *(National Archives)*

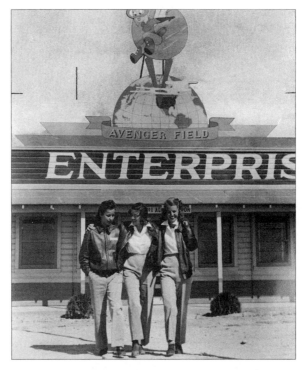

(*Above*) The field of aviation offered a variety of new job opportunities for uniformed women during World War II. One of the most unusual and exciting of these jobs was that of flying aircraft. Approximately 1,000 women wore the santiago-blue uniform of the Women Airforce Service Pilots, a quasi-military organization, and ferried aircraft of all types throughout the United States. (*Special Collections, Texas Woman's University, Denton*)
(*Below*) WASP trainees study map reading and navigation at Avenger Field, Sweetwater, Texas. (*Special Collections, Texas Woman's University, Denton*)

(*Above*) The wishing well at Avenger Field where WASP were tossed after they soloed. (*Special Collections, Texas Woman's University, Denton*)

(*Below*) WASP Wilda Winfield on a photographic mission at Frederick Army Air Forces Base, Oklahoma. (*Special Collections, Texas Woman's University, Denton*)

In this photograph WAVE Air Traffic Controllers direct aircraft arrivals and departures at the Naval Air Station, Anacostia, D.C. 1943. *(National Archives)*

A WAVE aviation metalsmith works in the assembly and repair department at the Naval Air Station, Jacksonville, Florida. July 1943. *(National Archives)*

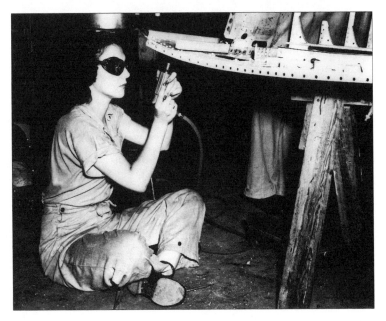

WAVE mechanics work on the engine of a Douglas R5D Aircraft at the Naval Air Station, Oakland, California. 1945. *(National Archives)*

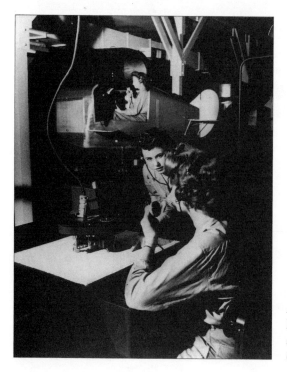

A Woman Marine at Cherry Point, North Carolina, operates a link trainer. 1943. *(National Archives)*

A Woman Marine works as a parachute-rigger at Cherry Point, North Carolina. November 1943. *(National Archives)*

Throughout the wartime years, mail was universally recognized to be the number one morale builder in the service person's life. Women and men in the military always looked forward to "mail call." WAAC Mary E. Blackman sorts the mail at Fort Knox, Kentucky. *(Women's Army Corps Museum)*

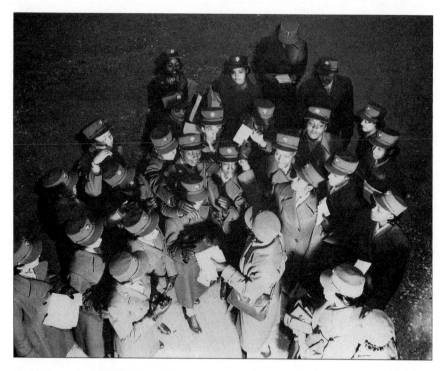

WAACs at Fort Huachuca, Arizona, a major military installation for African Americans, fall out for "mail call." December 1942. *(Women's Army Corps Museum)*

WAACs Ruth Wade and Lucille Mayo service a truck at Fort Huachuca. December 1942. *(National Archives)*

SPARs in foul-weather gear aboard the training vessel *DANMARK*. *(Historian's Office, U.S. Coast Guard)*

SPAR motor machinist's mate tests the battery on a truck. *(Historian's Office, U.S. Coast Guard)*

SPARs stand watch in a communications center during the war. *(Historian's Office, U.S. Coast Guard)*

SPAR ensigns armed with pistols prepare for a confidential mission for the communications office. *(Historian's Office, U.S. Coast Guard)*

Women Marines in December 1943. After completing boot camp at Camp Lejeune, North Carolina, she was trained as an aviation machinist mate at the Naval Air Technical Training Center in Norman, Oklahoma. In September 1944, she was sent to the assembly and repair shops at the Marine Corps Air Station in Cherry Point, North Carolina. At the end of the war, Private Turnbull was transferred to the Marine Corps Air Station in Ewa, Hawaii, where she worked on the flight operations line as a mechanic. Her letters were written to her parents and her sister, Jean. (For other letters by Turnbull, see pages 218–219.)

Camp Lejeune, North Carolina<br>January 1, 1944

Dear Folks:

. . . Right now, I have one ambition. That is to be a good Marine. This is the best organization in the world and I'm so proud to be a member of it. I'll continue to love it if they don't give me an office job. That will be about the worst deal in the world as far as I'm concerned. . . .

Love, Bub

Rather goodnight. I'm so damned tired I think I'll hit the sack.

Norman, Oklahoma<br>[March 1944]

Dear Jeanne:

. . . This week we are working in the shop. Last night we made patches in metal skins—the shape of an airplane fuselage. Making a waterproof aluminum patch is a lot of fun. Today we work on learning the fundamental parts of an engine. . . . Got to muster for class now.

Love, Bub Write.

Norman, Oklahoma<br>[March 19, 1944]

Dear Folks:

. . . Only twenty weeks of school left. Last week wasn't half bad. I don't know what my average was for the week, but it was ok. I don't know what kind of a rating I'll get out of this. The instructors don't think the women should be in this course so they smack down our grades so that the fellows get the rates. This week we work in the shop making fabric patches on airplanes and one thing we make is a dzus key which we use all through the course. It's used the same way a screw driver is only it's flat. . . .

Did I tell you that the course I'm getting is worth between $5,000 and $6,000 and would take two years elsewhere? . . .

Love, Bub

Betty Ann Farris of Detroit, Michigan, joined the WAVES on her twentieth birthday, in March 1945. After completing boot camp at the Hunter College Naval Training Station in the Bronx, New York, she was sent to the Naval Air Station at Olathe, Kansas, where she received training as a flight orderly. She was then stationed in Olathe where she was a crew member on Navy planes that carried military passengers to various locations throughout the United States. Occasionally, she served as a flight orderly on planes carrying the wounded to Stateside hospitals. Because she was only twenty when she joined the WAVES, she had to have her parents' written permission to enlist. She loved her work in the WAVES, and in a letter to her parents dated November 3, 1945, she exclaimed, "My ½ yr. anniversary in the Navy! Six whole months! It doesn't seem possible. I've loved every minute of it and have never once regretted joining so I'll never be able to thank you enough for signing those papers for me, Daddy. It has been the most wonderful experience of my life." Betty Ann Farris remained in the WAVES until her discharge in July 1946. Her letters to her parents include enthusiastic descriptions about her work as a flight orderly.

Olathe, Kansas<br>August 1, 1945

Dear Mother,

Well, I just got back this morning about 5 after a most wonderful trip. . . . There is so much to tell about the trip I don't know where to begin. Going to N.Y. we made stops at Chicago, Cleveland, and N.Y. and coming back we stopped at Washington, Columbus and Olathe. The cities were so pretty from the air, especially the capital building all lit up in Washington. The view of N.Y., the ocean and the ships in harbors was beautiful. The lights looked so pretty from the air too. Then all the farming country looked just like a patchwork quilt. It was so pretty, and such a thrill. We flew over the clouds and they looked just like big white soft pillows.

It's just like riding in a bus or train. I didn't feel the least bit sick at all and none of my passengers got sick either, thank goodness. Last night we ran into a lot of rain and thunder and lightning too and we were thrown around quite a bit.

The pilot and co-pilot were swell about explaining things. . . . I had to get coffee and sandwiches and etc. for them a lot and stand by while they landed to turn off the generator switches. . . . .

This is going to be quite a tiring job as it's pretty hard to sleep in the daytime in the summer, but I know I'm going to love it. . . . I'm going out again on Saturday and Sunday, I guess. Be writing you soon.

Love, Betty

Dear Mom,

Olathe, Kansas
August 2, 1945

. . . Well, I'm scheduled out on a hospital flight Saturday morning at 11:30 a.m. Hope too many aren't in stretchers. A doctor always goes along though. Don't know whether I'll go to Boston with them or to Tennessee. . . .

Love and Kisses, Betty

Dear Mother,

Olathe, Kansas
August 27, 1945

. . . Had a swell trip back yesterday. It's fun travelling by day. However, there was a terrific head wind and it was the roughest trip I've ever had. I darn near got sick. I went up to the cabin to heave (there was only an orange in my stomach anyway) and a Navy Commander Doctor there with the pilots talked me out of it. So I wouldn't spoil my record though I didn't eat a thing all day. I had several passengers sick, even the old salts.

The scenery is beautiful from Cleveland to Detroit and from there to Chicago as we go over Lake Erie and Lake Michigan. . . . Any trip now I expect to be put on the western run for about a month. . . .

Going to chow now. Bye. Love to all, Bet

Dear Mother and Daddy and Do Do—

Olathe, Kansas,
September 15, 1945

. . . I had the best time coming back—the swellest bunch of passengers. It was rough though and several got so sick. When they got off they thanked me for being so good to them. . . .

I don't know what the heck I'll do when I get out. I hate the thoughts of getting a discharge. I love this life so much. . . . I hate the thought of sitting behind a desk 8 hrs. a day after what I've been doing. I might try airline hostess, I might be just a farmer's daughter and help him, *or* I wouldn't mind getting married. . . .

This has turned out to be a long letter, hasn't it? Do you appreciate the time I put into them? I hope you're saving them all for me cause I want them as a diary.

Have to write some more now. Be good, write soon.

Love, Betty

One of the most unusual and exciting of the new jobs available for wartime women was that of ferrying aircraft across the United States for the Women Airforce Service Pilots, a quasi-military organization affiliated with the Army Air Forces. From September 1942 until December 1944, when the organization was disbanded after it was not accorded full military status, approximately 1,000 women had the distinction of flying military aircraft of all types throughout the United States for the WASP. Yet not until 1977 did the Women Airforce Service Pilots achieve full veterans' status. The WASP gloried in their work, and their wartime letters are filled with details of their love for flying.

Madge Rutherford of Indianapolis, Indiana, graduated from Butler University, where she learned to fly in a Civilian Pilot Training Program. She joined the WASP in January 1943. Following training at Avenger Field in Sweetwater, Texas, she was sent to Long Beach, California, where she ferried planes. On October 10, 1943, she married Sherman Minton, a Navy doctor stationed in San Diego. The couple had originally planned to marry the previous April, but Rutherford postponed the wedding after Jacqueline Cochran, the founder and director of the Women Airforce Service Pilots, wired her in January 1943 with instructions to report to Chicago if she wanted to fly in the WASP. In July 1944, when it became apparent that the WASP would not be granted military status, Madge Rutherford Minton resigned from her duties. After the war, Madge and Sherman Minton began the study of herpetology. They and their three daughters have traveled throughout the world collecting reptile specimens. In 1969, the Mintons co-authored *Venomous Reptiles*. When they are not traveling, the Mintons reside in Indianapolis.

[Sweetwater, Texas]

Dearest Mother and Dad:                                     April 3, 1943

. . . My snap rolls are all smoothed out now and I love them. Yesterday my instructor and I couldn't seem to gain more than 2500 ft. of altitude so we found us a quiet little corner and looped, rolled, and vertical-reversed from that height. If he wasn't such a superb pilot, I would never have permitted same! He told me not to spend too much time solo on acrobatics because my efforts were o.k., but to continue work on chandelles and lazy 8's. I ran another stage, 180 degrees this time, and glided too flat to suit the little men who graded me, otherwise o.k. I am going out this afternoon, if the weather clears, and glide correctly to redeem myself.

We are off to work—pounds and kinetic energy in physics now. I have about a 90% average which I can't understand since I feel constantly confused in class.

Let me again sing the praises of the situation. Honestly, I truly love it here. Remember "Loch Lomond"—here's the Avenger Field words to the chorus:

"You take the runway, and I'll take the mud-hole,

And I'll hit Avenger a'fore ye,

For me and my PT are standing on our nose,

In the muddy muddy field at Sweetwater.

The mud is abnormal, of course, but the gals sing lustily. We call ourselves the "Hut! 2– 3– 4 glamour girls." . . .

Love, Madge

[Sweetwater, Texas]

Dearest Mother and Dad:          April 7, 1943

We didn't fly today. The wind made a complete 360 degree circuit today, hitting all the cardinal points and has at last settled back to Southeast which means very unsettled weather here at Avenger.

Talk about solid comfort! After eating two, not just one, but two huge T-bone steaks, a salad, and a piece of chocolate pie, I have assembled my writing materials and have sought solitude in the front cockpit of PT # 132 [Primary Trainer] which is sitting on the side line with seat well-cushioned and adjusted. At last I have found a place to be alone in the wild island of feminine charms and can write letters in peace. It seems positively cozy and homelike. Goodness knows it should, considering the time I spend in one of these cockpits. . . .

Two more weeks only, after this week, and the Fairchild [a P.T. Airplane manufacturer] will be behind me. Looking thru my log-book I found that I had flown over 40 different P.T.s so far, all having their little eccentricities, I assure you.

The sunrise this morning was extra special lovely and I cannot but again extol the beauties of Texas generally—the broad-domed sky, gentle slopes, lovely winds and broad broad stretches of earth and more earth.

I wish you could have seen the landscape as I saw it yesterday from 6500 feet. . . .

Love, Madge


                                                          San Antonio, Texas
Dearest Mother and Dad:                                      May 11, 1943

Yesterday I dragged myself off to the Link Training building, since we did not fly, for an hour of sheer torture only to be most pleasantly surprised and to have my attitude toward that unpredictable little training almost completely changed. [The Link Trainer was a machine that simulated every sort of flying condition. Pilots spent much time in the machine as they prepared for every eventuality.] Link has been the bane of my existence from the second hour to now but I really liked it yesterday. Under the hood, I set her on a course of 90 degrees and the instructor turned on "rough air." I mean rough! It wasn't long before I was fighting rudder and stick with glee as the trainer pitched, yawed, and rolled, and I was holding my course within 3 degrees at all times which was pretty good. After 20 minutes of this, I made climbing, and gliding turns in rough air, trying to hold banks and proper rate of turn. It was glorious fun and I passed the exercise without difficulty. I shall not dread my next Link period now. . . .

Love, Madge


The following letter, written after Rutherford had completed her WASP training, describes a flight across the United States via the southern route.


                                                      [Long Beach, California]
[Dearest Mother and Dad],                                   March 2, 1944

. . . Pre-flight of ship is accomplished at the Vultee factory and I proceed to beat my knuckles raw on the empennage and wings of my assigned plane, checking cracked ribs, etc. You won't have one once in a hundred planes but they are moved about in pretty crowded quarters and you can't risk not discovering a defect before take-off. . . .

Take-off isn't so bad so you climb to 500 feet, 90 degree her to the left, 45 degree her to the right, and steady her down in a course of 75 degrees. After about 25 minutes your marker beacon light glows and you know that you are just west of the first mountain pass of the many to be navigated during the trip. It is San Jacinto, a gaunt igneous upheaval of 10,000 plus feet whose deep gullies are filled with hardy mountain pine now crushed under tons of snow and ice and whose sheer grey granite canyons and cliffs seem tremendously raw and wild to your Mid-west born and bred daughter.

But once through the pass you are suddenly 40 degrees warmer and wish you could shed your leather jacket . . . you are letting down to the Palm Springs Army Air Base where there is a pursuit school for ferry pilots. . . .

The next morning saw us on our way to Phoenix, 270 miles distant. The great Arizona desert is just that. It is rock and sand, mountains of both. Radio beams are uncertain, they bend and twist around the mountains and the country is destitute of railroads or highways. So the way you navigate is, after passing over the tip of the Salton Sea, you point your nose to Black Mesa and fly. . . . Phoenix is a gas stop and after a quick lunch at the Airline Terminal you fire up and head for Tucson. . . .

There are three passes out of Tucson and the choice is sometimes hard to make. . . . You cautiously steer around the mountainside and start for Cochise Head, a natural great stone face beside Cochise Pass. . . . Ahead is the Paso del Norte . . . and you dip your left wing around a huge statue of Christ on the mountain top as you cut a fine path between the low red-roofed huts of Ciudad Juarez and the towering roofs of El Paso. . . .

Morning finds you streaking toward Midland through Guadalupe Pass—the last mountain bastion east of the Rockies. Across Big Springs, Avenger Field with its neophytes, Sweetheart Lake with its mesa-topped hills, Abilene, Ft. Worth, you skid into Dallas just under the deadline of "on ground one hour before sunset" and hastily clear the runway for the B-24 landing on your tail.

Next afternoon, after spending the morning loading up the ship, sweeping the frost off the wings, and warming up a very cold engine, I took off for Monroe, La. . . . Next afternoon a weather check reported a clear to Greenville, and we . . . took off and threaded Old Man River up to our destination.

The tower here was very unhappy with us for the runways were barely cleared of snow drifts and were ice-glazed, but we had little trouble; a fact that completely demoralized base pilots who have been grounded for days because of the conditions of the runways and hated to see a woman beat their time. But they were courteous. Then, the papers are signed; you check your forms, and gather your luggage while the ground crew wheels the ship away to the hangar where they will completely dismantle it and put it back together the army way.

[Love, Madge]

Another WASP who delighted in her work as a wartime pilot was Yvonne "'Pat" Pateman of Sewaren, New Jersey. Following her WASP training at Avenger Field, she served as a ferry and test pilot at Romulus Army Air Field in Romulus, Michigan, and Shaw Army Air Field in Sumter, South Carolina. After the war, she accepted a commission in the Air Force, retiring as a lieutenant colonel in 1971. She has logged over 5,000 hours of flying. The pioneering efforts of Pat Pateman, as well as those of her WASP sisters, have helped pave the way for the successes of Sally Ride and other contemporary women in aviation. The letters that follow were written to Mary Ann "Jerry" Wetherby, a former classmate of Pateman's at Avenger Field, who was forced to withdraw from the WASP program because of illness in her family.[13]

Sweetwater, Texas<br>June 4, 1943

Jerry, Dear:

BTs [Basic Trainers]—oh, baby, there isn't anything like 'em. When it comes to this flying game, sister, you ain't even started to fly until you've flown a BT. And I'm not kidding! Even PTs [Primary Trainers] have to take a back seat with the little cubs. . . . Well, I'll stop this and start at the beginning. It really felt wonderful that first morning when we marched out to the flight line, and then turned abruptly and marched to Hangar 1 instead of the old hangar. That was when we all realized that we were finished with primary training. Then to go out and sit in this massive piece of machinery and be allowed to start it and to stop it. All the little gadgets hanging around, to make us mixed up. But as soon as a guy gets to know what each little one means, he goes at it automatically. But today—ummmmm—we flew a BT! Honestly what a sensation. The instructor tells you to go through the cockpit procedure, then call the tower—go down to No. 1 take off position—call the tower, take off and climb to 500 feet—well little girl, you just feel like you're in heaven— That great big nose sticking up in front of you, scares the daylights out of you—but you taxi nice and slow and "s" all the way because it's too big you don't want to run into a little old PT. And the first time you call the tower you get kind of scared that they won't hear your little voice. Then as you begin giving it soup to takeoff, it really becomes alive. (We takeoff on the runway and land to one side of it.) Torque correction, and drift is easily handled on runway takeoffs because you can see when you're

---

13. The Yvonne C. Pateman letters are located in the Yvonne C. Pateman Papers, United States Air Force Academy Library, USAF Academy, Colorado.

going off. Then it climbs. We get all bollixed up because the altimeter is set at sea level, 2,380 feet. But we get used to that too. It's really a big thing. This course will make pilots out of us or break us. Checking instruments is as important as holding your altitude. It's odd to be flying in the same area with ATs [Advanced Trainers]. You kind of have to look around a bit. It is an airplane that we're flying now! And you've got to be flying it every minute. . . .

Love from Texas, Pat

Sweetwater Texas
June 19, 1943

Jerry, dear—Sittin' in my bay

Got your letter—thanks—well, well at last—at LAST!!! I soloed to-day—made three landings—and it was wonderful. Simply wonderful. Checking everything—calling tower—"FF81 from 106 in No. 1 position ready for immediate takeoff. Over." The tower: "FF81 to 106— you are clear to takeoff. Over." Me (little voice) "106 Roger." Then give it the soup and really run up the power—We takeoff on runway at 500 make level turn, climb to 600, turn on to downwind leg, then while turning on baseleg call tower again—"FF81 from 106 turning on base leg—requests landing instruction over." Tower: "106. You are No. 3 to land west of the runway (or on the runway, etc.). Watch your spacing. Over." And then me "106 Roger." Then you come in and hold her off until you're real close—she lands beautifully—all that weight— And now it's all over and everybody is so happy. Everyone looks up to more solo—cross country, etc. . . .

   Must call this off, Kiddo, please write soon.

Love, Pat


As war casualties mounted, hospitals and rehabilitation centers were enlarged and new facilities were established to care for the wounded. The American Red Cross provided a variety of support services for these establishments. By the end of the war, 50,000 Red Cross women worked as civilians in uniform at Army and Navy hospitals. The following letter was written by Harriet Raab, a member of the Arts and Skills Corps of the Red Cross. She worked at the Navy Hospital in Farragut, Idaho.[14]

---

14. Foster Rhea Dulles, *The American Red Cross: A History* (New York: Harper & Brothers, 1950), 382–83. The letter of Harriet Raab is located in The Harriet Raab Papers, U.S. Navy Collection, World War II Survey, United States Army Military History Institute, Carlisle Barracks, Pennsylvania.

[Farragut, Idaho]
Dear Ones: August 24, 1943

I want to give you all a summary of what things are like here at the naval hospital in Farragut. We are stationed in the Civilian employees dorm. There is no room in the nurses' quarters. The rooms are adequate, though not fancy—single bed, dresser, little bedside table, desk table, big chair. . . . We eat in the officers' and nurses' mess. It is cafeteria style. . . .

This hospital is very highly organized, and on the whole very efficient, if I can judge. Because of the intricate organization and the shortage of materials, it is very hard to get things. I have an office given to me in the physio-therapy dept. It is not large enough, but a lot better than nothing. . . . That won't be nearly room enough, and the navy will squawk when my place isn't neat, but I can't help it.

Now, as to the program. I had thought until today that it would be just in the wards, but there is a large navy auditorium here, and we can have shows in that, using corpsmen (male nurses), patients, and others. The captain who is the executive officer wants us to have a minstrel show and use the auditorium. It is the strategic thing to do to have the show, for it will allow us to have the aud. It will be difficult to have such a big thing right at the start, but I think that it can be done. My chief problem will be to find people who are talented in that type of entertainment—as you know, that has not been exactly in my line.

The ward program will consist of craft work, such as carving, modelling, square knotting, sketching, making Christmas cards, woodburning, waffle weaving, braiding. Fortunately, I know all of these crafts, many of which I have learned in the past 2 weeks. Then we will have games, parties such as birthday parties, talent shows, informal music, etc.

The girls here in the Red Cross are very nice. I think that I am going to enjoy it here, but it will be a lot of work, as you can imagine. I am 35 miles from the nearest town, which is too bad, but this place is very complete, and that makes it not too bad. . . .

[Harriet]

The next letter was written by Loretta Carney of Chicago, Illinois, who served as an Army dietitian at the 2,000-bed Veterans Administration Hospital in Danville, Illinois. Carney's letter is addressed to her sister, Catherine Smith of Green Bay, Wisconsin.[15]

---

15. The letter of Loretta Carney is located in the Horace Smith Papers, Area Research Center, University of Wisconsin–Green Bay Library, Green Bay, Wisconsin.

Danville, Illinois

Dear Catherine:            October 22, 1943

. . . We are very busy here. The hosp. has been enlarged by 400 beds to take care of the casualties from this war, of which we have some 200 or more, many are back from Africa, Guadalcanal, and other battle fronts. There are 3 dieticians here for 2400 patients, but, the next to the chief is expecting a transfer any day now, so that will only leave the chief and I; Four have been allowed but there aren't any to be had. I have the Hosp. division alone, some 700 patients; there are some 76 bldgs on the grounds and the other patients are in some of them; only my sick ones are in the Hosp. bldg. The help situation is very bad. Out of a dept. of 96 male employees 44 are already in the Service and more to be called soon, all replaced by women, which is far from satisfactory. In spite of the fact that it is a Govt installation, the food situation is grave, of course, meats, sugar etc. all canned goods, rationed, just as on the outside, and we scarcely know from day to day what kind, if any, meat will be available. Menu planning is just plain H---. . . .

Love, Loretta

When they were not on duty, women in uniform often led active social lives. They spent weekend leaves visiting nearby towns and cities, went to USO shows, danced to wartime swing music at local nightclubs, saw scores of movies, ate restaurant meals away from their duty stations, and enjoyed a variety of activities, such as roller skating, swimming, bowling, and playing tennis. From reading the letters of women who were stationed at Stateside postings, one almost gets the sense that social activities were available every night of the week. One young WAVE, expressing the opinion of many of her contemporaries, wrote to her parents and assured them that even though she might "talk about seeing so many different fellows and having dates with so many," she was not "fickle." Later, she wrote, "I have met hundreds of fellows since I've been in service. Some older, some younger, some richer, some just like ourselves, some with high-school educations, some with college degrees, some handsome, some ordinary. . . ." But she reassured her parents, "I choose my companions very carefully."[16]

As one might expect, some of these military romances resulted in marriage. Despite the widespread misconception that women in the military could not be married, marriage did not disqualify women from

16. Eunice G. McConnell to her parents, June 27, July 7 & 27, 1943.

enlisting, nor was it grounds for discharge. Certain restrictions against marrying did apply, however, and commanding officers at local bases sometimes exercised their discretionary judgements.

Ensign Janet Mary "Mike" Murray of Johnstown, Pennsylvania, a member of the Navy Nurse Corps, met her future husband, Lieutenant Robert Wooddall of Fairburn, Georgia, while they were both on active duty. As with many war marriages, the exigencies of the time required that Ensign Murray introduce herself to her future parents-in-law by letter.

U.S. Naval Hospital
St. Albans, New York
May 2, 1944

Dear Mr. and Mrs. Wooddall:

This is a pretty difficult letter to write but I have wanted to do it for a long time.

I only wish it were possible for us to meet, because a letter seems so empty, but under the circumstances there is no alternative.

Perhaps you'll both be able to come to Florida, for the wedding. I sincerely hope so. . . .

I realize how you both must feel about his marrying a girl whom you have never met. My own parents feel the very same way. Golly, I'm not doing so well in writing this, am I? But, it all boils down to just this. —I love Bob, very much and always will, and I know that with God's help we will have a very happy married life. . . .

I do so want to meet you, because I want you both to be satisfied with Bob's choice. I'll do my best to keep you from ever worrying about his welfare. . . .

Sincerely, Mike

Allan Berube has demonstrated in *Coming Out Under Fire: The History of Gay Men and Women in World War Two* (1990) that thousands of homosexuals served in the United States military during World War II. The following letter, written by WAVE Mary Liskow, studying to be an aviation machinist mate at the Naval Air Technical Training Station in Norman, Oklahoma, offers an insightful glimpse into the types of problems that lesbians in the military encountered.[17]

---

17. The Mary Liskow letters are located in the Mary Liskow Papers, Bentley Historical Library, University of Michigan, Ann Arbor, Michigan.

[Norman, Oklahoma]
Dear Folks: January 3, 1944

Sorry I haven't had time to write but everyone in the barracks has been upset this week. The whole affair is very unfortunate but had to come out sooner or later. It seems there has been a case of mental homosexuality or so I'll call it, and people didn't understand what it was. Both of the girls involved are swell kids in fact one is more or less on the genius side. She has been a company commander and very active in all the sport events. . . .

Some of the girls who sleep near them have reported it on various occasions although I have seen nothing out of the way myself. The situation has been growing more serious for the last month until a group of the girls went to the chaplain for advice and he took it higher. This, of course, divides the barracks into two groups. One who thought they were angels and martyrs and those who thought they were doomed. About a month ago this same group of girls warned them that they would have to stop and that if they didn't they would be reported. It finally boiled down to the fact that it was taken out of their hands and the girls were discharged New Year's Day. I person- ally think what happened was that V., who had just heard her fiance was killed in action the day she left to join the WAVES, was lonesome and F. took advantage of the situation and it got out of hand. The barracks has been so divided about the whole affair that it practically started another Civil War between the North and the South. The thing is gradually simmering down and I hope will eventually burn out. I do feel sorry for the kids but there's no getting around the fact that any- one who plays with fire will get burned sooner or later. The latest report is that they are both going to their own homes where I hope they can get some help to straighten it all out as they are both well worth the effort. Perhaps it's better, as the ensign said, to "keep our dirty linen at home," but what happened all the time was that it was kept under cover so long it became almost too late. Of course it will be just as well that this doesn't go any further, but I thought you might be interested. . . .

Love, Mary

Service in the military brought people together from many differ- ent ethnic, religious, geographic, and socio-economic walks of life. Bernice Sains Freid, who served in the WAVES during the war, re- cently observed, "I had never been to a museum, seen an ocean, a mountain or a waterfall. Therefore, when I entered the Navy, I was like a WAVE in Wonderland."[18] In letters to her friends back home in St. Paul, Minnesota, she wrote with awe about her travels throughout the United States. Yet, as the only Jewish servicewoman in Yeoman School at the United States Naval Training Station in Stillwater, Oklahoma, she sometimes felt very alone and isolated.

Stillwater, Oklahoma<br>
December 5, 1944

Dear Marion:

I love it here in Oklahoma. I'm on a campus, and I feel like I'm in college, which I wanted to attend but had to work. . . .

Something happened that I wouldn't dare tell my parents. After I was here 2 weeks, I suddenly felt very lonely, a real melancholy swept over me that I couldn't shake off. This went on all day. Suddenly I realized what was bothering me. It was Friday night. It wasn't a rabbi or a synagogue I needed. It hit me that I was the only Jew on the campus, and there are 200 of us. I didn't dare tell my superior officer; she wouldn't understand. I felt I couldn't tell my 2 room mates, nice Christian girls, who would think I was peculiar. I couldn't wait until the next day, Saturday, when we would be off after lunch til Sunday night. I decided to go to Oklahoma City, where I certainly would find Jewish people.

I was a mess that Saturday. I stood up on the 2-hour ride to Okla. City, didn't speak to anyone. When we arrived, I jumped from the bus like a crazy person, asked someone blindly for a USO. Fortunately, it was only 2 blocks away. I flew there. In front of the USO I had to collect myself, walked in like a person dying of thirst. Luckily, to my left was a huge blue banner with the yellow letters J.W.B. [Jewish Welfare Board, one of the six organizations to establish the USO]. I felt like I was being saved, like a dying man in the desert at an oasis. My heart pounded; my eyes were filled with tears. I managed to walk in, trying to act normal. There on the bench was the homeliest skinniest little sailor I had ever seen. At any other time I would have ignored him, but I knew he was Jewish, and at that moment, that was all that mattered. I sat down next to him. I don't remember what I said to him, but gradually I calmed down.

Then the J.W.B. man in charge came in. I asked him if there were other Jewish service personnel in the area. I'll never forget his reply, "There are 30-40 Jewish men at Ft. Norman, Okla., but you're the ONLY JEWISH SERVICEWOMAN we've ever seen."

I was cured instantly. Every Sat. night a Jewish family entertains

18. Bernice Sains Freid to the authors, [September 1992].

us at their home. Of course I attend with fellows. We sing songs around their piano mostly. The family has a son in the Army. Then we are driven back to the USO. . . .

I'll be sorry to leave Stillwater. I've made some nice friends here. We've been asked to select billets for duty assignments. I pray I'll get San Francisco. I wouldn't dare ask for a naval air station. I don't dare take a chance on being the only Jewish person in a place again. . . .

Love, Bernice

Military service also enabled African Americans to travel to distant places, meet new people, take on new jobs, and confront new challenges. For both black women and men in the military, however, this almost always occurred within segregated units. Of the original 440 women who reported to Fort Des Moines, Iowa, for WAAC Officer Candidate training in July 1942, thirty-nine were black. These women were housed separately, assigned separate seats in classrooms, and required to eat at tables especially reserved for "Colored." This practice of segregation continued throughout the wartime years for WAAC/WAC enlisted personnel and officers. Moreover, the fact that thirty-nine of the first 440 officer candidates at Fort Des Moines were black was no accident. The War Department had established a quota that stipulated that up to 10.6 percent of the WAAC/WAC could be black, but the Army never came close to reaching this number. At peak strength, early in 1945, about four thousand black women were in the WAC, a number that amounted to about four percent of the Corps.[19]

Black women were not accepted into the WAVES and the SPARs until November 1944, a change that only occurred after substantial effort and protest by progressive organizations. Near the end of the war, in July 1945, the Navy reported that there were two black officers and seventy-two black enlisted women in the WAVES. The Coast Guard reported that four black women had been accepted into the SPARs. Both black WAVES and SPARs were fully integrated into their service branches. Neither the Women Marines nor the WASP accepted African Americans, however.

When the war ended, approximately 500 black nurses were serving in the Army Nurse Corps. These nurses mainly cared for black soldiers and prisoners of war. More black nurses would have enrolled in the

---

19. Treadwell, *Women's Army Corps,* 596. Martha S. Putney, *When the Nation Was in Need: Blacks in the Women's Army Corps During World War II* (Metuchen, N.J.: Scarecrow Press, 1992), 1–2.

ANC had it not been for the implementation of a quota system during the early years of the war. Not until January 1945 were black nurses admitted to the Navy Nurse Corps. Only four black women served in the Navy Nurse Corps, and they were fully integrated into the Navy. Two hundred and fifty-two black women served with the overseas division of the Red Cross during the war years on a segregated basis.[20]

Despite the segregation and discrimination experienced by blacks in the military, the African-American community demonstrated strong support for the ideals for which World War II was fought. A. Philip Randolph's March on Washington Movement used as its motto, "Winning Democracy for the Negro Is Winning the War for Democracy." The "Double V" campaign, adopted by the black press in March 1942, called for victory over totalitarianism abroad and racism at home. It was in this context that significant numbers of black women volunteered for military service.

The following three letters were written by black women requesting information about joining the Army Nurse Corps. They were sent to Judge William H. Hastie, Civilian Aide to Secretary of War Henry L. Stimson. Hastie, a former dean of the Howard University School of Law, was responsible for issues concerning the fair treatment of African Americans in the military. These letters exemplify the desire and willingness of African-American women to come to the aid of their country in its time of great need.[21]


Judge William H. Hastie, Civilian Aide
Secretary of War                                        Athens, Georgia
Washington, D.C.                                     December 11, 1941
Sir:

I am urged and moved by the hour of trial which has visited our country to offer and volunteer my services either as a nurse, clerk or social worker or in any capacity which might require a combination of

---

20. Jesse J. Johnson, *Black Women in the Armed Forces, 1941–1974* (Hampton, Va.: Hampton Institute, 1974), 6, 33, 48. Robin J. Thomason, *The Coast Guard & the Women's Reserve in World War II* (Coast Guard Historical Office, 1992), 5. Kathryn Richardson Tyler, "The History of the American National Red Cross, Volume XXXII, American Red Cross Negro Personnel in World War II, 1942–1946" (Washington, D.C.: American National Red Cross, 1950), p. 16.

21. These letters are located at the National Archives and Record Administration (hereafter cited as NARA), Record Group 107, Entry 91, Secretary of War, Office, Assistant Secretary of War, Civilian Aide to the Secretary, Box 225.

these services. I have had experience in all these fields. I am ready and anxious to serve in any branch of the government service, and would be willing to serve anywhere in or out of the country.

In the other World War, I served as a Regular Army Nurse, having volunteered and enlisted shortly after graduation from the People's Hospital of New York City. So anxious was I to serve that I put my age up to the minimum requirement. I have maintained and am actively registered as a nurse in the State of New York. I have done some social work and for six years worked with an insurance company as a general office clerk.

I shall be happy to receive a command whenever any need for my services become desirable.

Respectfully yours, Mrs. Mary C. Harris

The next letter was forwarded to Judge Hastie by Colonel Campbell Johnson, executive assistant to the Director of Selective Service.

Wilmington, Delaware
Colonel Johnson, [June 30, 1942]

I am writing for some information as to what I am supposed to do or where I am to go to find out where I can learn to be an army nurse. I am a colored girl twenty years old. I completed two years of Childs Nursing in Balti., Maryland. This course consisted of some First Aid, Obstetrics, and Physiology plus everything regarding the care of a child. I would be interested in institutional wards if you could give me some information regarding this. I thank you very much.

Respectfully, Margaret Anderson

Judge William H. Hastie
Civilian Aide to the Secretary of War Fredericksburg, Virginia
Washington, D.C. July 27, 1942

Dear Sir:

I was advised by Mr. P. B. Young of the Journal and Guide Publishing Co., Inc. of Norfolk, Va. to write you for information which I desire.

I am a young woman, aged 34. I have done hospital work, and if it is possible with your help, I would like to enter a Nurses Training Course for Defense Work.

I did not finish high school, but with the experience I've had, I can assure you I will do my best and will not disappoint you should

I be able to get some help from you, say, for instance, a letter of some sort.

I worked in Riverside Hospital, North Brothers Island, N.Y. doing nurse's aide. I also did Practical Work (Residential) for W.P.A. [Works Progress Administration] in N.Y.C. for a while. . . .

I am, yours truly, (Mrs.) Pearl E. Tibbs

Coping with prejudice and discrimination was a regular part of the experience of black women in the Army Nurse Corps. Letters written by black Army nurses to Mabel K. Staupers, Executive Secretary of the National Association of Colored Graduate Nurses (NACGN), often described these deleterious conditions. Staupers served as the unofficial watchdog to highly placed government and military officials during the war years. As the executive secretary of NACGN, she chivvied public officials, helped organize letter-writing and telegram campaigns, and issued statements in behalf of the integration of black nurses into the military. The first letter was written by Mabel Staupers to Truman Gibson, who, in 1943, succeeded Hastie as Stimson's Civilian Aide. The letter requested information about the status of black nurses in the Army.[22]

New York, New York<br>September 20, 1944

Dear Mr. Gibson:

I would like very much if you would send me the following information concerning Negro nurses in the Army Nurse Corps:

The present number of nurses in the Corps—

Where are these nurses stationed both in Continental United States and Overseas—If it is not possible to give the location of the Overseas Unit, we shall be glad to know what Theatre of Operations they are in—

We would like to know how many nurses have been promoted to the rank of Captain and First Lieutenant. If it is possible to have the names of these people it would be most helpful in writing our future history.

I would also like to know whether there are any discharges and retirements.

We understand that Cadet nurses are now serving in Army hospi-

---

22. The next eight letters are located at NARA, Record Group 107, Entry 91, Secretary of War, Office, Assistant Secretary of War, Civilian Aide to the Secretary, Boxes 225, 228.

tals and we are wondering if any Negro nurses have been assigned to these hospitals and where.

Is there any truth in the rumor that the hospital at Huachuca [Fort Huachuca, Arizona] is to be closed, if so I am wondering what will happen to the nurses who are there—

With kind personal regards,
Sincerely yours,
Mabel K. Staupers, R.N.

The next letter, written to Staupers by a black Army nurse, describes the anger, frustration, and bad morale that resulted from the discriminatory practices of the ANC. The letter writer draws attention to the fact that black Army nurses were sometimes required to perform menial jobs in place of and in addition to their nursing assignments. She also describes how the common practice of assigning black Army nurses to care for German prisoners of war was "a 'bitter pill' to have to swallow."

Station Hospital
P/W [Prisoner of War] Camp
Papago Park, Phoenix, Arizona
October 29, 1944

Dear Mrs. Staupers:

. . . For five months we have had to do all our scrubbing and cleaning of quarters, while the white male and P/W officers have someone to do theirs. We are the only women on the entire post and no help whatsoever. We are told that no P/W or enlisted man can work in women's quarters. The white nurses before us certainly had someone to clean for them. To do what we are and have been doing is against Army regulations.

Apparently we are not considered officers by those in command, for we are never included in the command affairs and meetings called for all officers of the post to attend.

Last night there was a large reception given at the Officer's Club to welcome the new post commander. It was a command affair and every officer besides being urged to attend, was given an invitation. That is, all the officers on the post but the five Negro nurses. As time goes by conditions seem to get worse than better.

. . . There is some kind of recreation on the post for every one but us. Phoenix is a little "south" in itself. We cannot be served in any restaurant, Kreese's, Woolworth's or soda fountain because of our color. Yet, the war department seems to think Phoenix should offer an outlet to our isolation problem. . . . Thus every night week after week, and month after month we sit in and stare at each other. It is not a normal life for anyone, after spending eight and a half hours with the Germans every day. . . .

I just received a letter from one of the nurses who is with the 168th Station Hospital in England. They feel pretty much let down to learn they travelled all that distance to take care of German prisoners. It is really a "bitter pill" to have to swallow.

. . . The nurses in India seem to be getting along alright so far. At least they are not taking care of prisoners. They too, have three white captains over them.

Even tho Capt. Petty left here as chief of our group, she has ended up over there, under white heads, as surgery supervisor. I imagine that was a let down to her.

Well, I could go on and on relating parts of letters from our girls over there, but it only lowers my morale here.

Sincerely yours, [Name unknown]

Government officials received a number of inquiries from concerned citizens requesting information about the treatment of blacks in the military. In the following letter, Major Edna B. Groppe, ANC, assigned to the Office of the Surgeon General in Washington, D.C., provides an official explanation of Army policy toward black nurses. The letter was written to Frances M. Williams of the National Federation for Constitutional Liberties and is reprinted exactly as it was written.

                                          Office of the Surgeon General
                                          Washington, D.C.
Dear Miss Williams:                        November 6, 1944

. . . There are now eight negro Army nurses receiving basic training at Camp McCoy, Wisconsin. This training is taking place at an Army installation at which both colored and white nurses are assigned. At present there are 276 negro nurses commissioned as officers in the Army of the United States. During the first world war there were no negro nurses in the Army. After armistice eighteen were assigned who had previously been in the process of being considered.

The plans for increasing the number of nurses do not discriminate between colored and white nurses. Generally nurses are obtained through application to the American Red Cross. However, applications for commission as officers to serve in the Army Nurse Corps may be made directly to the Surgeon General's Office. . . .

The negro nurses when taken into the Army become a part of the integrated negro and white program. Negro nurses are not segregated from white nurses in the basic training period. They attend the same classes, drills and other parts of the training program, and eat at the same mess. Negro nurses, however, are housed together as a matter for their own convenience. When the basic training is completed negro nurses, the same as white nurses, are transferred to their posts within the continental limits of the United States and to overseas assignments as the needs arise. When in the Army negro nurses are subject to all the rules and regulations of the Army in the same manner as white nurses and are subject to the same orders and the same disciplines. Neither the present program nor any future plans are conducted along segregated or discriminatory lines.

The qualifications for admission into the service for negro nurses are the same as for white. New applications will be given fair and impartial consideration and their assignments when in the Army will be in accord with the changing needs as they develop in various foreign theaters and upon this continent.

Very Truly Yours, Edna B. Groppe, Major, A.N.C.

Black women were always a part of the Women's Army Corps. Yet a system of racial segregation and discrimination prevailed in the WAC throughout the wartime years. Mary McLeod Bethune, a prominent black educator who served as Director of the Division of Negro Affairs for the National Youth Administration during the New Deal years, acted as a special advisor to Colonel Hobby on questions relating to black women in the Women's Army Corps. Harriet M. West, an assistant to Bethune during the 1930s, was a member of the "Pioneer 440" at Fort Des Moines, and she was stationed at WAC Headquarters in Washington where she served as an advisor on questions relating to black women in the Corps. In the next letter, WAC recruiter Dovey M. Johnson writes to West and discusses some of the problems she faced in recruiting black WACs. Johnson, who was also a member of the first WAAC officer regiment, was known at Fort Des Moines as "the walking NAACP."[23] After the war, she had a very distinguished legal career.

---

23. Putney, *When the Nation Was in Need,* 141.

Cleveland, Ohio

My dear Major West:                                    May 13, 1944

I regret that I shall not be able to see you before you leave for Adjutant General's School. Today, as during that time—which seems so many years ago—when we were O.C.s [officer candidates] together you have been a source of quiet strength to me. I think that thousands of Brown American WACs could say the same to you and I don't think I shall ever forget the masterful speech you made in Columbus at my recruiting meeting. You'll never know how often I recall the things you said then and take new courage for The Next Hour's Task. . . .

Here is my problem.

1. I sent you, I believe, the intelligence that here in the Columbus District Recruiting Area a special day has been set aside for processing Negro applicants.

2. The necessity of this has been explained on a basis that Cleveland hotels approached on the subject of quartering our girls who volunteer to work for democracy refused them.

3. Today notice has been given that from now on Negro girls will be processed in Columbus and other than Negro girls in Cleveland. (I believe I told you that our headquarters moved from Columbus to Cleveland.)

4. In Columbus quarters processing for all WAC applicants was on a democratic basis.

5. These things are of great concern, because in this area our people are aware and fear the army. They hesitate to join the WAC. They mistrust the principles for which the Army stands. The community, as in every other community, is keenly sensitive of democratic malpractices accorded our men and women in the Army.

6. These facts clearly show that this practice will hinder recruiting seriously. A nationwide WAC recruiting program was initiated today, May 13. It is my firm belief that the success of the coming Invasion will depend in large measure upon the complete and absolute unity of this drive as it concerns every American irrespective of race.

Your objective analyzation of all situations suggest that I write this to you. . . .

Respectfully yours, Dovey

In the following letter, Major West reports to Truman Gibson on the transportation problems she encountered while touring southern military installations. Black service personnel traveling in the south routinely experienced incidents such as the ones described by Major West.

SPWA 330.14 (22 Feb 44) E                    22 February 1944
Subject: Transportation difficulties encountered in the South.
Through: Director, WAC
To: Mr. Truman K. Gibson
    Civilian Aide to the Secretary of War
    Washington, D.C.

1. In accordance with verbal request from the Civilian Aide to the Secretary of War to the Executive Office of the Director, WAC, the following report is submitted:

a. On 22 January 1944, the undersigned left Washington, D.C. for Camp Forrest, Tennessee to join Mrs. Lula Garrett Patterson, Women's Editor of the Afro-American newspaper. Other than the necessity for riding in a dirty half-coach from Chattanooga, Tennessee to Tullahoma, Tennessee, no difficulties arose to and from Camp Forrest.

b. On 25 January 1944, Mrs. Patterson and the undersigned arrived in Anniston, Alabama. Upon leaving the train at Anniston, the undersigned asked the conductor (there being no porter in sight) to lift a regular size suitcase from the train. He remarked, "I don't handle baggage for Niggers." Inasmuch as the trip from Fort McClellan, Alabama (Anniston) to Fort Benning, Georgia was made by private car, no further incident occurred at this point.

c. On 28 January 1944, Mrs. Patterson and the undersigned drove with two WAC officers from Fort Benning to Columbus, Georgia. Mrs. Patterson and the two WAC officers went into the station to purchase tickets from Columbus to Atlanta. The undersigned went directly to the train to wait for them, standing near the last coach of the train while waiting for the others. A man in workman's clothes, and a cap with a visor approached, saying, "Your coach is down there back of the engine." The undersigned did not answer. He then said, "Hey, you, don't you hear me talking to you." Still no answer. Whereupon he grabbed the undersigned by the arm, turning her around, and called her several *very* vile names, using most *obscene* language, and saying that when Niggers were in that section they respected white people, or words to that effect. The undersigned jerked away saying, "Take your filthy hands away from me, or you won't be able to put them on any other woman," and "Why don't you join the Nazis, that's where you and your kind belong." The man then said he was going to get the police and put the undersigned in jail. He was told to go ahead, whereupon he left going to the station, presumably to get the police. In the meantime, the WAC officers and Mrs. Patterson came out and Mrs. Patterson and the undersigned boarded the train. The coach was filthy, the covers on the seats were black and grimy. So much so, it was necessary to put kleenex over the seat before sitting down. No further difficulty ensued.

d. On the morning of 29 January 1944, 0925, aboard the Seaboard train #10, Mrs. Patterson and the undersigned asked the porter if it was too late for breakfast. He explained that if we were in the diner within a half an hour we could be served. Upon arriving in the diner we were ushered to a section behind curtains, where four pullman car porters and two waiters were having breakfast. There were two vacant seats opposite the two waiters. The steward advised us that that section was for Niggers and that was where we would eat. We refused and returned to our car.

2. It was the understanding of the undersigned that even though separate facilities were provided on trains in the south, they should be equal. It was found that no equal facilities are provided on coaches, on any southern railroad or in the diners on the Seaboard Railroad.

Harriet M. West
Major, WAC


The following letter, written by a WAC stationed at Camp Forrest, Tennessee, was published in *The Houston Informer* on May 26, 1944, under the headline JIM CROW SWEEPS SOUTHERN CAMP IN AN OPEN LETTER TO THE PUBLIC. The editor of the newspaper, Carter Wesley, also sent a copy of the letter to Truman Gibson. The irony that the events described in this letter occurred at Camp Nathan Bedford Forrest, which was named after the Confederate general who founded the Ku Klux Klan in 1866, did not escape many observers. Throughout World War II, Camp Forrest had the reputation of being one of the worst duty stations for black military personnel.


Dear People:

Camp Forrest, Tennessee
[May 1945]

> Did not our forefathers fight and die
> For these United States.
> Have we not answered every cry
> To make this nation great?

I am a Negro in Service. I am doing my part for this country. My home is in Chicago, formerly of Texas. I have been in the WAC since WACdom. I am really disheartened more than words can say, that is, I, and thousands of decent girls, abandoned colleges, clubs and friends who were most dear to us to join the Women's Army Corps to do a job and do it well, feeling that we were needed and could do our share to help the boys over there. Upon entering the Women's Army Corps, we were told by WAC officials that we would be given equity, opportunity to exhibit our skill—that is, the qualified ones would have professional jobs, such as dieticians, stenographers, librarians, etc. whereas the less skilled but qualified ones would have the opportunity to go to Army specialist school and complete the course work which was most beneficial. But under no circumstances was it stated that we would be sent to a place like this terrific Camp Forrest, Tenn. and be tossed about like a chip on a sandy shore by *white civilians* working on this post.

The only types of jobs here for the educated Negroes (WAC) to do toward victory are: Nurses' Aides (doing all the dirty work), pushing food carts, washing windows, woodwork, scrubbing floors and cooking in the hospital messes, taking orders and being cursed at by civilians working there.

Here if a civilian yells and curses at a WAC and tells her to do something, regardless of what it is, she either executes his orders or is subject to court martial.

Recently a white civilian went to the extreme to strike (slap) a colored WAC and endeavored to pour hot grease on her in one of the messes. And what do you suppose this Tennessee white man got as a punishment? His punishment was a two day lay off with pay and a return to his job at camp. Such drastic occurrences cause more damage than the Japs and Germans put together.

Thus I feel that we have made a great sacrifice to join the fight for democracy. We don't want the whites to pity us, but we are true Americans. Our boys are across the sea fighting and dying for this country and the things we all think are worth fighting for, while over here at home and various camps insanity sweeps the South.

I'm sure that all the white people in the South don't approve of such drastic occurrences. It is just the few that are trying to keep America down. America is supposed to be a free country, but without the aid of Negroes it is a doomed country!!!!!

Yours Truly,
Constance E. Nelson, WAC

Please print it all. I didn't have time to type it, but please try to understand my writing. It is vital that the people know about their Colored WACS.  Thank you.

The *Chicago Defender* also received letters that called attention to discriminatory practices in the military. The following letter, written by a WAC stationed at Fort Des Moines, Iowa, was sent to John H. Sengstacke, publisher of the *Defender*. The author of the letter drew poignant conclusions about racism in the United States military during World War II.


                                                        Fort Des Moines, [Iowa]
Dear Mr. Sengstacke:                                    January 8, 1944

As a member of the Women's Army Corps, I am deeply conscious of this war against fascism and have dedicated myself to do all possible to bring the day of victory closer. In order to win a war against fascist ideas, I feel strongly that the same ideas at home must be combatted, that we need the best that everyone has in them as individuals, that a discriminatory racial policy can only interfere with the winning of the war. For this reason, therefore, I am writing to you, asking that you lend your voice in support of the Negro WAC.

President Roosevelt has said that millions of men and women are going overseas during the next year. We take that to mean that Negro WACS as well as white WACS may be sent across, and, because of our confused policy concerning Negro men overseas in the Armed Forces, we in the WAC would like to know what the policy is going to be toward the Negro WACS overseas. Are they going to be sent as part of a segregated unit to be used to do the work of a labor battalion or will they be sent as part of a democratic unit and placed in jobs according to their training and ability?

At Camp Pickett, Va. and Camp Forrest, Tenn., the first is true; they are part of what could be termed a labor detachment. The women are not placed according to their abilities. With such things happening right here in the United States, we want to be certain it will not be duplicated overseas.

The Army Specialist Training Program at Rutgers; Grinnell, Ia; New York University and City College, NYC, has no segregation, no discrimination. The men are learning a living lesson in democracy. Is the WAC to be accorded less?

These are the questions the WACS are asking—the answer must be one of victory over fascist ideas of racism. Knowing your devotion to the cause of freedom and equality for all, and your earnest desire to do all you can to win victory quickly, I can only feel that you will do all within your power to help remedy this situation which can only cause disunity among the American people and help to prolong the war.

Sincerely yours [name withheld]

The next letter was written by WAC Private Nellie R. Holiday, a light-skinned African American, who was originally assigned to a white unit. Her letter describes the specific problems this situation created for her.


Truman K. Gibson
Washington, D.C.                                    Chicago, Illinois
Dear Sir:                                          February 26, 194[5]

I am a WAC stationed at Florence Army Air Field, in Florence, South Carolina, at present I am A.W.O.L. and will really kill myself before I'll return to that, excuse the expression, "hell's hole."

In March, 1943, I was sent to Ft. Oglethorpe, Ga., for basic Tng. After four weeks I was made P.F.C. and was sent to Dening, New Mexico. I was placed in a white outfit, then I asked to be transferred to a colored tng. unit because I am very much a Negro despite my make-up. I was told by the commanding officer there that the social standing of the Negro was so low that I would not want to be associated with them. . . .

When the WAAC changed to the WAC I took a discharge and stayed out almost a year, but it is more or less a tradition in my family for some member to be associated with the armed forces. I reenlisted at the recruiting station in Chicago. I stated my problem and received assurance that I would be with a colored outfit . . . yet I was shipped to a white outfit. I was treated royal, a swell job and what have you, but it is more than I can take. The cracks, being associated with people whose only thought is to keep the Negro back. I can not have my boy friend who has been overseas for nearly three years return to me wherever I am. I can't have my father and mother visit me, yet I can't get a transfer. My problem is too much for me. I am not hiding out. I'm staying at the same address I enlisted from and am asking you to offer me your aid. I do not wish to return to S.C. and I do wish to return to the war. *With* a *colored outfit* where I *belong heart and soul.* Please lend me your advice and aid. I'll be waiting for your answer at the above address.

Sincerely,
Pvt. Nellie R. Holiday


On February 28, 1945, Gibson's office sent a telegram to Private Holiday that advised her to turn herself in to the provost marshall.

After following these instructions, Holiday was transferred to a black unit in Sioux City, Iowa.

Black women were not accepted into the WAVES and the SPARs until November 1944. What follows is a special message to black women calling for them to join in the campaign to end the ban against blacks in the WAVES. It was written in July 1943 by Thomasina Walker Johnson, the legislative representative for the influential black sorority, Alpha Kappa Alpha.[24]

National Non-Partisan Council on Public Affairs
of
Alpha Kappa Alpha Sorority
961 Florida Avenue,
Washington, D.C.

TAKE A LICK AT THE ENEMIES ON THE HOME FRONT
TAKE A SHOT FOR DEMOCRACY AT HOME

After trying for nearly a year to get Negro women admitted to the Auxiliaries of the Navy, we were told about 5 weeks ago that a plan for the admission of Negro women was under consideration and was waiting for the approval of Secretary [of the Navy, Frank] Knox. Efforts to discuss this plan with officials of the Navy proved fruitless. On Saturday, July 10th, we learned from reliable sources that the plan had been approved but that the announcement of it had been withheld until they could find a school at which Negro women could be trained and, further, that the plan did not provide for Negro women officers. We have taken the following position.

1. We want Negro women admitted to the auxiliaries of the Navy on the same basis as other women—in training, work, chance for advancement, etc. (At present all American women citizens are permitted in these services—Chinese, Indian, etc., except Negro.)

2. We are tired of "steps in the right direction." (For us it is always the final step). We want this time, simply, to have what every other American woman citizen has already. *This is the only step.*

3. We want no compromises!

WHAT YOU CAN DO: (And if you don't do it, when Negro women are given an inferior status, don't grumble. YOU are to blame. With your help and cooperation we have a chance to gain our first victory on the Home Front!)

<hr/>

24. The Thomasina Walker Johnson letter is located in the National Association for the Advancement of Colored People Papers, Group II, B. Legal Files, Library of Congress, Manuscript Division, Washington, D.C.

1. Send a letter IMMEDIATELY (or telegram, card, etc.,) to President Roosevelt and to Secretary Frank Knox saying, simply, that you are opposed to any plan that is different in ANY way to that now in effect for any other American woman citizen.

2. Get on the phone and get as many other persons to do the same thing as you can possibly get!

3. Your Congressmen and Senators are now at home. Call them up and ask them to write or telegraph Secretary Knox voicing his protest.

4. If for any reason (job, position, etc.) you cannot write a protest—get others to do it. WE MUST NOT LOSE THIS BATTLE FOR DEMOCRACY!

Sincerely yours,

Thomasina Walker Johnson, Legislative Representative

In late 1944, when the Navy finally began to admit black women, it did so on an integrated basis. The following letter, written by a black WAVE in boot camp at the Hunter College Naval Training Station in the Bronx, New York, was addressed to Walter White, the Executive Secretary of the National Association for the Advancement of Colored People.[25]

Bronx, New York

Dear Mr. White,                                                    [1945]

Just a few lines to inform you of the goings on between the Navy (Women's Reserve) and the colored recruits.

We mingle as freely with each other as though we were all of the same race. Everyone has the same opportunities and we may go up as far as we like.

I honestly enjoy being a WAVE, and am looking forward to many interesting, happy, and worthwhile experiences while in the Navy.

I know that you, as Secretary to the N.A.A.C.P. are interested in all progressive steps of the colored race.

To hear it from a person who is experiencing it is not only cause for gratitude, it's an authentic report of what is really going on.

Sincerely yours,

Clementine B. Forsyth, A. S. [Apprentice Seaman]

---

25. The Clementine B. Forsythe letter is located in the National Association for the Advancement of Colored People Papers, Group II, B. Legal Files, Library of Congress, Manuscript Division, Washington, D.C.

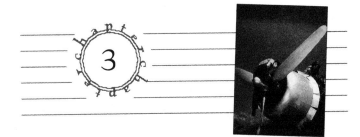

# Stateside—Zone of the Interior

THE OVERWHELMING MAJORITY OF United States women who donned uniforms during World War II served at Stateside or zone-of-the interior billets. While WACs, Army and Navy nurses, and Red Cross women were assigned to overseas posts throughout the world during the wartime years, they were only a small minority of uniformed women. At the end of the war, 17,000 WACs, 35,000 Army and Navy nurses, and 7,000 Red Cross women were stationed outside the continental United States. In addition, a few WAVES, Women Marines, and SPARs, all of whom had originally been prohibited from serving outside the continental United States, received assignments in Hawaii and Alaska during the final months of the war. Yet even the women who served outside the continental United States usually received their basic training, schooling, and advanced training at Stateside locations.[1]

This chapter focuses on letters written by five uniformed women who served in the United States during World War II. These letters, from members of the WAC, WAVES, SPARs, Women Marines, and WASP, help illuminate the meaning of the wartime experience for women who were assigned to Stateside duty.

---

1. Mattie E, Treadwell, *United States Army in World War II, Special Studies, the Women's Army Corps* (Washington D.C.: Department of the Army, 1954), 772. "The Nurses' Contribution to American Victory," *American Journal of Nursing* 45 (September 1945): 683. Foster Rhea Dulles, *The American Red Cross: A History* (New York: Harper & Brothers, 1950), 374.

KATHERINE "KAY" TRICKEY OF Lewiston, Maine, graduated from the University of Maine in 1932 at the height of the Depression. Over the next few years she held a variety of jobs. When Dow Air Force Base opened in Bangor, Maine, early in World War II, she qualified for and obtained a good clerical position at the installation. In the autumn of 1943, she enlisted in the Women's Army Corps with a group of Maine women who took the oath of allegiance to the service in a ceremony at the State House in Augusta as a part of a special recruiting effort. These women underwent basic training together at the Third WAC Training Center in Fort Oglethorpe, Georgia, before being assigned to separate duty stations. Katherine Trickey wrote regularly to her parents in Maine. Her basic-training letters are replete with information about the regimen and routine of Army life.

Portland, Maine<br>November 16, 1943

Dear Mother,

I hated to say good-bye, but don't let yourself feel too badly because I shall be doing what I *want* to do. I'll be seeing new places and doing different things.

The ride down was glorious. In the light from the bus, the trees on the edge of the road shone bright and glistening. The evergreens were loaded with snow and even the deciduous trees were coated on the bare branches. The fields were white and often I could catch a glimpse of a snow covered country road. The lights in the houses shining out on the snow covered lawns made cheerful spots in the dark landscape. There was snow even in [Portland]. There couldn't have been a nicer farewell ride between Lewiston and Portland unless it had been in the summer. I was glad it was snowy before I left. . . .

Much, much love, Kay

Fort Oglethorpe, Georgia<br>November 28, 1943

Dear Folks:

I can't remember when I wrote last. . . . Saturday we had our first formal inspection, so Friday night we had a *GI* party [Service people refer to thorough cleanings as GI parties] with scrub brushes, dust cloths, etc. trying to make our barracks the cleanest of all. Friday nights are reserved for *GI* parties! Saturday noon I got *five letters*. I know now how the boys feel when they have mail call and no letters. Those letters Sat. sure pepped me up. . . .

We had small-pox vaccinations and typhoid and tetanus inoculations Sat. afternoon. I got through it all right *without fainting*. It wasn't bad at all. My arms are slightly sore today, but really not too bad. . . .

My but I'm glad I'm here. I think I shall enjoy practically every minute of it during training at least. Even the work is fun when all of us are doing it together. . . . We march everywhere, to mess, to classes, to work (even to the movies until we get our uniforms). . . .

Love to all, Kay

Fort Oglethorpe, Georgia
December 13, 1943

Dear Folks:

We are attending a class on First Aid. This past week has been so busy. A week ago yesterday I had KP all day so I didn't get a chance to write any letters and I haven't been able to get caught up yet. Monday when we got home from classes in the afternoon we found that our daily inspection had been so bad that we were ordered to spend the evening cleaning the barracks. Tuesday evening I had a big washing to do and it went like that all week. Classes this week have been interesting. We are taking up Manners for Military Women, Interior Guard Duties, Uniform Regulations and Care of Clothing, Military Sanitation, Preventative Medicine, Backgrounds of the War, Articles of War, Company Administration, Physical Training, and Drill. We've had some very interesting films. . . .

I am still getting quite a thrill out of being here particularly as we march to class mornings and nights with the band playing and company after company march through the trees to classes. . . .

Loads of love to all of you, Kay

After completing basic training early in 1944, Katherine Trickey was transferred to Camp Wheeler, Georgia, just outside Macon, for the remainder of the war. Camp Wheeler was named for Confederate General Joseph Wheeler, and it served as a major training facility throughout World War II. More than 218,000 soldiers received some portion of their training at this facility. At Camp Wheeler, Private Trickey worked as a typist and file clerk.

Dear Folks,                                    Camp Wheeler, Georgia
                                               February 23, 1944

I am slightly late on my letter this week. I have been alone in the files as my corporal is taking training this week out on the firing range. . . .

I had Saturday afternoon off and went in town hoping to get my hair done but every place was full. . . . After supper, we went up to the USO and read a while and then watched some soldiers doing fingerpainting. We could have tried it too but neither of us were in the mood so we just watched. The hostess at the USO was very cordial and nice. We enjoy going there. . . .

My corporal is out again today training so I must stop this and get to work. If I do the work without any indexes as he has been doing it, it only takes about three hours a day to put everything away—but it takes hours every time I want to look up anything. I imagine I'll make myself work and index most everything just so I can find it quicker. I hate hunting through paper after paper and telegram after telegram. Yesterday I hunted through three months of incoming telegrams which took me about three hours and if they had been indexed I could have done it in half an hour easy. But it takes plenty long to index them. I started on January's yesterday and worked about four hours and only got half done. However, if it is done each day, it isn't too bad a job, once I have caught up-to-date on it.

Love, Kay

Katherine Trickey, as did many women in the service, took good advantage of the opportunities to travel and visit new areas as well as to meet new people. As a Maine native stationed in Georgia, she was particularly aware of the distinctive southern atmosphere of her assignment. She took several trips to Atlanta where she walked down Peachtree Street and visited the famous Atlanta Cyclorama, a large circular painting dating from the late nineteenth century, which depicts the Battle of Atlanta. She commented on the warmth of the southern hospitality she and other WACs encountered during a trip to the University of Georgia at Athens, and she used a three-day pass to visit St. Augustine, Florida. The sweltering heat of Georgia's summers did not go unnoticed. During one particularly hot spell in August, she wrote about a "watermelon party in the barracks day room" in which "we spread papers all over the floor and about a dozen of us crowded around. *Messy* oh my, juice and seeds and rinds all over that paper! . . . More fun."[2] The following two letters focus on Katherine Trickey's travels in the South.

Camp Wheeler, Georgia<br>March 6, 1944

Dear Folks,

You have heard, undoubtedly, of Georgia as the Peach State. Yesterday, I found out why. Six of us hired a car and drove to Fort Valley to see the peach blossoms. It was the Sunday when they were in full bloom and we had been told it was a sight worth seeing. It certainly was. There are acres of trees. The pink blossoms against a background of a heavenly blue sky was a sight I shall never forget. We got out of the car and walked a long ways through one orchard. We got so far in it that we could see nothing but peach trees in any direction. . . . We couldn't have had a better day. The weather was hot and summery but not uncomfortably so. It seemed grand to drive a car again. The OPA [Office of Price Administration, in charge of rationing and price control] allows them to rent one car for 90 miles pleasure driving per individual per month. It seemed queer to have to watch the mileage so carefully. As there were six of us in the car it cost us only 2.10 a piece and we were gone over eight hours and drove the full 90 miles. . . .

After we had eaten dinner at Fort Valley in a small but nice restaurant there, we walked around the town to see the houses and flowers. . . . We asked an elderly man who was out in front of one of the homes about one of his flowers and got into a conversation with him. He finally invited us into his house to meet his wife and we had quite an interesting visit with them. It was the first time any of us had been in a civilian home since we joined the Army. . . .

It was as you can see a very interesting day, much of it made more interesting through the friendliness of the Georgian people to those of us in service. We find that we have many experiences of this nature which would never happen to us as civilians. I don't know that we would be very interested in entering strange men's homes if we weren't in uniform!! . . .

Love, Kay

---

2. Katherine Trickey to her parents, August 7, 1944.

Camp Wheeler, Georgia
Dear Folks,                                    March 29, 1944

. . . Marjorie [WAC Marjorie Crockett of Caribou, Maine, also stationed at Camp Wheeler] and I had a very nice week-end in Atlanta. We got there about 6 o'clock Friday night and had a room at the Piedmont Hotel. Very nice with private bath. We especially enjoyed our baths in a tub instead of a shower. We slept late each morning and ate when we were hungry. . . . Saturday we shopped. . . . The city is queerly laid out. There are many business streets and . . . you have to hunt for your stores. . . .

Sunday we went sightseeing. We took a trolley out through the residential districts as far as Emory College. It is a very very attractive city in many ways. Of course as in so many southern cities in order to get to the white residential districts you have to go through the poor Negro sections which are nearer downtown but after you do get out the streets are very nice. . . .

Must stop now.

Loads of love, Kay


With 16.3 million service personnel and 15 million civilians on the move during the wartime years, direct connections through telephones and telegrams were often slow and cumbersome. Telephone companies exhorted civilians to avoid making telephone calls between 7:00 and 9:00 p.m. because that was the only time that service personnel could call "the home folks." Nonetheless, long waits at telephone booths often occurred. Yet most service people with Stateside assignments good-naturedly endured this inconvenience.


Camp Wheeler, Georgia
Dear Mother,                                   April 12, 1944

I guess my telephone call is not going to get through after all. I don't seem to have much luck with them, do I? I'm sorry. I had a class last night at 7 o'clock so couldn't start trying until 9. Then the operator said it would be 5 or 6 hours before the call would go through. There was a big crowd at the telephone building. Everyone decided that after Easter would be a quiet time to telephone, I guess. I went back to the barracks and put in a call for this morning but it hasn't come in and it's most time to go to work. I hope you didn't have too bad an evening sitting up for the call. I really didn't think I would be so busy or I never would have got your hopes up like that. Anyway, here is my belated Birthday Greetings, *dear*. . . .

Love, *dear*, Kay

Camp Wheeler, Georgia
Mother, Mae, Phil and Family,                    May 2, 1944

A hasty note to let you know that I got another promotion yesterday. You may now address me as Sergeant Trickey. However, when you write my title is Tec 4 Katherine W. Trickey. These Technicians ratings are queer. A tec 4 is called a sergeant and gets the same pay but is lower in rank than a Sgt. ($78 per month minus bonds, insurance and laundry.) I am, of course, very pleased to get it. Marjorie got Tec 5 (or Corporal) for which I am very pleased. She has been very jealous of my rating and I am afraid is still jealous that I have gone ahead of her again.

I had a wonderful day Sunday just not doing anything in particular. I got up for breakfast then went back to bed for another nap, read awhile, sewed awhile, then ate dinner. Played baseball for two hours in the afternoon, gorgeous weather, sunny but a breeze. Went for a long walk in the woods after supper before dark. Then just talked with the girls and went to bed at 10:30. Peaceful and nice.

I finished my Projector course last night and have a license to run the 16 mm projectors. It has been interesting. I hope something else comes along for a good class.

Very hastily. Will write later again. Kay

Visits by dignitaries always created a storm of activity in the barracks and on military posts. When Colonel Oveta Culp Hobby, Director of the WAC, came to visit Camp Wheeler, the preparations were especially intense.

Camp Wheeler, Georgia
Dear Folks,                               May 26, 1944

This is a big day here. Colonel Hobby herself is coming to inspect us. My, have we worked! Last night was a great scrubbing bee. The barracks are really clean even if cement floors, beaverboard walls, and wooden beams can't be made to shine very much. My detail had to put oil on all four of our stoves and shine them; rather messy but didn't take too long. Then of course our own areas had to be done. Floor scrubbed with turpentine to take the tar off and then scrubbed on our hands and knees with soap and water and a brush. We had to make nice white beds which had to have an exact six inch fold and be exactly six inches between the edge of the pillow and the fold of the upper sheet. All our shoes had to be shined and I've got some new polish which really does the trick. . . .

May 27, 1944

We had quite a day yesterday. Having G.I.'d the night before we had only the last minute details to attend to. We got out of work at 3 o'clock and went to the barracks and changed into fresh uniforms and then just had to stay inside the barracks and wait. We couldn't sit down even for fear of wrinkling our skirts. At 5:15 Col. Hobby arrived. We had a formal inspection in the barracks; then we went to the mess hall where she gave us a very nice talk. I guess we all fell in love with her. She was very good looking, had a very sweet low voice, and an extremely pleasant personality. . . .

Much love, Kay


Camp Wheeler, Georgia
Dear Folks,                                    October 4, 1944

. . . Friday evening we went to the USO camp show at the gymnasium. Fairly good,— Rather good magician's act—and one of the dancing acts was good. Several comic acts which were good. . . . Sunday, I slept all the morning; then played tennis on our new court for almost 2½ hours. . . . I am much surprised to think I can hit the balls at all; it has been so long since I have played. I've done much better than I expected to, believe me. I only hope I can keep it up as well as I have begun. It's been heaps of fun to play again. I needed the exercise badly and am getting it all in a dose!! . . .

Love to all, Kay


In the next letter, Katherine Trickey describes the new responsibilities that came with a job promotion. The challenges that she now faced as the supervisor of other clerks were the subject of several letters to her parents.


Camp Wheeler, Georgia
Dear Folks,                                    October 19, 1944

. . . As for my new job—it is an expansion of my old one really as I have both the decimal file and the 201 [standard Army personnel files]—Minnie is able to do the decimal file herself and likes it better now that I don't so closely supervise her work! The other girls are ok. . . . P. is a young rather uncultured girl who isn't bad to work with although she is always in trouble about something as she is very scatterbrained and has no sense of right, wrong or of responsibility apparently. She is likeable, however.

M. is harder to get along with. She is loud and overbearing. She has had one year of college and is probably quite capable but she won't put her mind on anything except every body else's business and her evening dates. . . .

Much love to all of you, Kay

Camp Wheeler, Georgia
Dear Folks,                                          November 8, 1944

. . . The election news is in and our friend Roosevelt won again. I really expected him to do so, but [Thomas E.] Dewey made a better showing than they expected, I guess. Maybe it's just as well at this particular time, but I still think we're crazy to give one man that much power. I don't trust his peace policies either. I'm afraid of what he will commit us to be without our knowledge and consent. Maybe I'm just pessimistic and maybe he really does have our, not his interests at heart.

Love to you all including the youngster, Kay

November 9, 1944

We did go to Macon and practiced basketball. It seems good to be taking part in athletics again. Thought I was too old for that sort of thing, till I joined the WACs!!

Throughout the wartime years, travel on buses and trains was often very crowded and difficult. This was especially true during holiday seasons. The following letter describes a tedious train trip Katherine Trickey experienced during a 1944 Christmas furlough.

Camp Wheeler, Georgia
Dear Folks:                                          December 31, 1944

. . . The trip to Washington was o.k. Not exciting and no one to talk to, but I read and slept and enjoyed myself. From Washington down, it was a beastly trip. I didn't get a seat until morning. We got on the train at 11:35 p.m. and it didn't even leave the station until 1:30, so it was two hours late in starting and it stopped every little ways and just crawled anyway so that it was four hours late reaching Atlanta. The train was just as crowded as it was in June. Soldiers and sailors sleeping in every spot of the floor and even one in our car who slept on the baggage rack.

Fortunately there was another WAC on my car who offered to take turns with me in her seat which was very decent of her. So she, I and a soldier took turns about an hour at a time. . . . Of course, I missed my bus to Macon and had to buy a train ticket. That train was also late so I reached Camp just an hour late. The girls who made up the Morning Reports fixed it up for me as if I'd gotten in on time, however, apparently with the approval of the officers so I haven't heard any more from it. . . .

Loads of love to all, Kay

Katherine Trickey served in the segregated military. Moreover, she was stationed in the deep South, where segregation was a dominant way of life. Consequently, she had only limited contact with African Americans. Her letters contain occasional references to the social and economic hardships experienced by southern blacks. Yet she felt that these hardships were more rooted in economic than in racial circumstances. After a bus trip through Fort Hill, the black section of Macon, she wrote the following letter to her parents.

Camp Wheeler, Georgia

Dear Folks,                                               April 29, 1945

. . . We took a bus that said Fort Hill and it turned out to be a ride practically entirely through the Negro section. Honestly, I don't understand why there are not horrible epidemics here in the South. The conditions as far as sanitation is concerned is appalling in the poor and Negro sections. The houses we saw today for the most part had no screens at all either in the doors or the windows. The yards are cluttered and many had cows and chickens in out buildings near the house. The toilet facilities are obviously the outhouse variety and this in a city that has running water. And in even the best sections of town there are alley ways of Negro homes that are nothing but shacks.

The poorer class of whites live in homes almost as bad. Many of the presumably nice homes are little better than our summer cottages! The wealthier homes, however, are lovely. . . .

Love to all, Kay

In the spring of 1945, one of the most disquieting events experienced by American citizens was the sudden death of President Franklin Roosevelt on April 12, 1945. Katherine Trickey's response was more cautious than many, but it does reflect her immense feeling of loss.


Camp Wheeler, Georgia
Dear Folks,                                        April 14, 1945

. . . We have all been rather stunned at the news of the President's death. It was so unexpected to most of us, although, of course, his last newsreel pictures showed his condition. There were memorial parades for some of the battalions this morning and short memorial services near the offices. Many of the girls are quite broken up over the news as they really worshipped him. I *do* feel that this is a very unfortunate thing to have happened right now, but, of course, I believe that there are other Americans capable of running the country. . . .

Love, Kay


With the dropping of the atomic bombs in early August 1945, the end of the war came quickly. Most military personnel marked the events with thoughts of going home, as well as with thanks that they had been spared from appearing on the casualty lists.


Camp Wheeler, Georgia
Dear Folks,                                       August 14, 1945

This is certainly one great day. It doesn't seem possible that it is true. I was in the movies in Macon when the President's announcement came tonight. I was alone and although I had just gone in—the movie no longer interested me, [and] I took the first bus back to camp. Macon was beginning to be noisy when I left it. People were standing in front of the stores under the awnings (it was raining hard!)—and all of the cars and buses were blowing their horns.

About halfway back to camp the sun came out and there was a gorgeous rainbow which seemed so fitting. About 9 o'clock, Lt. Kennedy came round and invited us to raid the mess hall for an impromptu snack. . . . I just wish it meant we'd be going home soon—but I'm afraid it will be Easter *at least* before we're through. There are many many to be discharged and someone is going to have to do all the paper work. I expect we're the ones who are going to do it.

Of course, the war is over for the civilians—not for the Armed Forces! All the broadcasts tonight kept emphasizing—2 day holidays for whom—the *civilians*. Military personnel report to work as *usual!!!*

*Who won the war, anyway—* . . . oh, well we should worry, so long
as it is over.

Much love to *all* of you, Kay


At the end of the Second World War, Katherine Trickey began to
give serious consideration to her postwar plans. Like many World War
II veterans, she intended furthering her education with the use of
funds from the G.I. Bill.


Camp Wheeler, Georgia

Dear Folks,                                      October 10, 1945

. . . I'm ready to move. It seems queer with people leaving all the
time. Something like the Junior year at school with the Seniors leaving!! One of the girls who got out in June couldn't stay away and has
been down here visiting for the last two weeks. Living in the barracks
and even helping at her old job in the office. Seems almost unbelievable, but tis true!!

I'm still as vague as ever as to what I want to do when I get out. I
wrote Simmons Library School [in Boston] and got a catalogue—
could do that next year maybe if I could find some way to live on fifty
dollars a month. Tuition would be covered o.k. . . .

Love, Kay


Fort Dix, New Jersey

Dear Folks,                                     November 30, 1945

Well, this is it. I arrived at Ft. Dix at 3:00 yesterday afternoon and am
scheduled to leave tomorrow afternoon. . . . It still doesn't seem possible.

Mother, I hope you won't mind if I spend a couple of weeks getting home. There are several things I may never get a chance to do
again, and some of the girls who have already moved from Cp.
Wheeler whom I want to see. I expect now to get home 2 weeks from
today for sure. . . .

Love, Kay


Following her discharge from the WAC, Katherine Trickey returned to Maine, where she used the G.I. Bill to earn two master's
degrees, one in teaching and one in library science. She worked for a
short time as a librarian at an annex of the University of Maine. She

then spent the next twenty-three years working in the Swampscott, Massachusetts, school system, retiring as Head of Media Services for that school district. After retiring, she returned to Maine to live. She has worked extensively as a volunteer for the Hampden, Maine, Historical Society.

TWENTY-ONE-YEAR-OLD Eunice G. McConnell of Brooklyn, New York, enlisted in the WAVES in February 1943. Following boot camp at the Naval Training Station at the Indiana University in Bloomington, she attended Storekeeper's School in Bloomington. In June 1943, she was posted to the Naval Air Station in Alameda, California, where she worked as a storekeeper. Her enthusiasm for the WAVES, her satisfaction in knowing that she was making an important contribution to the winning of the war, and her growing sense of self are recurring themes in the many letters she wrote to her parents.

Dear Folks,                                          Bloomington, Indiana
                                                     February 12, 1943

Well, I guess you're just about recovering from the fact that I am really gone. . . . It's 1750 now (that's Navy time, and on the Central Standard Time basis) and we've only been here since 0130, but nevertheless, I feel as though I've been in the Navy for years. If we keep learning things at the same rate we have done today—I ought to know enough to fill a 20 volume set of books. However, we probably won't.

First of all *you* will have to learn NAVY time too because I will probably always use it when I mention time. From 1 a.m. to 12 noon the hours are 0100 to 1200. From 1 p.m. to 12 midnight the hours are 1300 and so on up to 2400. An easy way is to subtract 12 from the first two digits if they are over 12—and the result is your p.m. hour. As—1815 (evening mess tonight) subtract 12 from 18, and you get 6— so it's really 6:15 that we'll eat. . . .

We signed in, and were sent to the dispensary for a slight physical exam—teeth, and hair was examined for lice. The BIG exam comes next Monday, I believe. Then we went to the mess hall and received chow on a metal tray that has 6 individual sections on it. You see we eat right from the tray arrangement—no dishes! We had a sort of meat loaf made from corned beef; a celery and tomato gelatin on lettuce;

peach pie and all the bread and butter and milk we wanted! (We were told that what we had was only left overs, so we expect better tonight.)

Next our bed linen and towels were issued. And, of course, we had to learn how a bed is PROPERLY made—at all times!

That may not seem like much—but I left out telling you how *many* times we had to go up and down the 2 ladders [as staircases are called in the Navy] leading to our deck, then the entire length of the deck, etc. Take it from me—it was quite a bit of walking!

We also received a mimeographed list of NAVY phrases we must know (at least by tomorrow) and instructions about use of the Head (bathroom) etc. All of this happened before I began this letter—what happens next, besides going to mess, we don't know! . . .

All My Love, Eunice xxxxxxx

P.S. So far all the girls are swell—we feel as though we've known each other for years!


                                                    Bloomington, Indiana
Dear Folks,                                         February 22, 1943

Thanks a million for the candy! Everybody liked it, even me. Today, as usual, we had a hard and long day. But being a part of the *Navy* we're all proud to do everything we can. As we are constantly reminded here—there is a WAR going on—and the fellows at the fronts have to keep on the go 24 hours a day—so we're not up to them yet! . . .

Love and kisses, Eunice xxxxxxx


                                                    Bloomington, Indiana
Dear Mom and Dad,                                   February 25, 1943

Gee, it was swell to hear from you, again. I say "again" because mail seems so slow and I just received your letter of Monday. Your letter sounds like an impossibility. I thought I had told you everything—but I guess I didn't. . . . I guess I told so many people about how wonderful it is here that when I got to you I must have figured that I had already told you too. Forgive me if I haven't and I'll try and tell you now.

First of all I love it. I thought I loved it before I got here, I *knew* it when I got here, and I'm *more sure* of it every minute that I stay! The University of Indiana is a beautiful place and our ship is a wonderful part of it. . . .

About the needles and X-ray, everybody gets them—no. I'm no special character! Yesterday we had a pelvic examination—gosh, it was embarrassing! Tomorrow we're to have more shots. . . .

Well, tomorrow is the BIG day! The uniforms will arrive and we are to try everything on to be sure they fit. Saturday is our official day to go into them, and we're going to have our first Dress Parade Captain's Inspection! That'll mean standing at attention (not batting an eyelash) for about ½ an hour. . . .

I'm not homesick (I hope that's really not a disappointment to you) but I really miss each and every one of you. . . .

Love and Kisses, Eunice xxxxxx

Bloomington, Indiana

Dear Mom and Dad and Eileen: April 27, 1943

Well, this letter will be packed with good news so you ought to enjoy it. I can assure you I'm more than enjoying it myself as I write it.

Your little girl is going to be a very changed person when you see her next time. The small changes have probably been taking place every day, but I didn't notice it at all until today. Hold onto your hats 'cause you'll never believe this one! Today I went to the Navy dentist, got there about 10:20 a.m. By 11:15 a.m. he had pulled a wisdom tooth and filled four others. I got up, walked out and returned to class. You know the old saying, "business as usual."

Then to top it all off we received our grades for last Friday's exams while at our last morning class, and I did very well! My average in all subjects came to 3.6!! [on a 4.0 scale]

Scandal: We have a girl in the brig. Charges: 1. She married during the training period and this contrary to regulations. 2. She married an officer—and it's almost a sacrilege for an officer to condescend and marry one of the members of the enlisted personnel. 3. She brought the officer aboard our ship (for about five minutes), and never logged him in at the Officer-of-the-Deck office. Expected results: She will probably be discharged dishonorably. . . .

Love and big hugs and kisses to you all, Eunice

After completing boot camp, McConnell was granted a short leave home. She then reported to duty at the Naval Air Station in Alameda, California. The next letter describes her cross-country train trip from her home in Brooklyn to her new assignment in Alameda.

Alameda, California

Dear Mom and Dad and Everybody, June 8, 1943

Well, here is the letter you have probably been waiting patiently to receive. . . . To begin at the beginning, my ride by coach from New

York to Indianapolis was made very comfortable and interesting by a sailor, also SK3c, (as myself) [Storekeeper Third Class] and who was stationed in Cleveland. . . . This sailor was from Conn. and was about 45 years old. He took care of me as if I were his daughter—treated me to dinner (costing about $1.75) in the diner, bought me an ice cream and later coffee, and sandwiches, etc. As it grew darker, he bought me a pillow, and persuaded me to sleep. All in all he was just one swell shipmate. . . .

Our train to Chicago arrived late—and we arrived at Chicago about 9:15 p.m. instead of 6:00 p.m. We tried to claim our Pullman reservations but had quite a bit of difficulty at first until I spoke to the Station Master and Train Master. About 20 minutes before train time we got straightened out. . . .

Now comes the most UNUSUAL part! We were accidently put on a Pullman car with Army Air Corps Corp'ls travelling to Sacramento! There were 12 Cpls and 1 Navy Warrant Officer with us 8 girls. They were surprised too and the Train Conductor nearly burst a blood vessel! We were the only girls on a train with about four Pullman cars and 17 coaches full of soldiers. . . . They were all very swell fellows. . . .

Kisses by the million, Eunice

Alameda, California<br>June 20, 1943

Dear Mom and Dad,

. . . Oh, Mom and Dad, I'm so happy. I'd be happy no matter where the Navy would send me. I'm happy because I know I'm doing my share to help our side win and at the same time making such unusual acquaintances and having such wonderful experiences. I suppose you can't understand how I can be happy and be so far from home and those I love. I miss you all—and think of you every minute of each day. But, on the other hand, I know you don't want me to be unhappy, so I'm taking advantage of all the opportunities that present themselves. You can be sure I choose my friends wisely—and after consideration. I shall always continue to use good judgment. In fact, I never bother to even keep a conversation going with any of the regular run-of-the-mill types. I stop so often and say to myself, "something must be wrong—I never had so much attention before. . . ." But, I just keep attracting the fellows, it seems. (I sound awfully boastful, don't I—sorry.) . . .

Just think when I'm home for good, I'll be one of those unwanted War Veterans who is always willing to chew off a monkey's ear with tales of what happened while in service. I guess I'm even doing that now—and I'm not a veteran yet.

All my love and kisses, Eunice xxxxxxx


Alameda, California

Dear Mom and Dad,                                    July 2, 1943

Well, here I am again ready to go to bellow off about how wonderful everything is out here.

My job seems well underway now, and I'm to be in charge of all the Stock Records of supplies sent to the Outlying Fields. Quite a responsibility, but that's what makes it really worth my time and effort. . . .

Your Baby, Eunice xxxxxxxxxxxxxxxxxx


Alameda, California

Dear Mom and Dad,                                 August 4, 1943

. . . It's 8:15 p.m. sundown, and the Flag is being lowered right outside the office windows while "retreat" is being played on the bugle. I'm so glad to be in service to think I'm actually a part of all this pomp and ceremony—as well as part of the backbone on which the boys at the front rely. . . .

Your Baby, Eunice


Alameda, California

Dear Mom and Dad,                              September 1, 1943

. . . Honestly, living here at the barracks with all these girls is really the most wonderful thing that ever happened to me. You'll probably find I've changed. It certainly teaches a person tolerance, and consideration for others and so many worth-while things. . . .

All my love and kisses for everybody, Eunice


Alameda, California

Hello Everybody—                                 October 6, 1943

. . . I'm rather sorry to hear the WAVES you see are sloppy—and those dirty hats, etc. Things like that give the public a degraded opinion of us all. But you may recall when I was at Training school I kept wishing I'd be sent to a Station where we would live in barracks and be under constant supervision. In this way the girls have to stand weekly inspections, etc., and are much more careful of their appearance at all times. We are always just according to regulations—short hair, clean neat appearance and all. It certainly gives your own personal spirits and morale a lift to hear civilians and others say as we pass by "Don't they look nice." And anyway, out here—each of us has so much pride about herself and her uniform that there's never any need for us to be reminded of personal appearance!

Now, you said something about me being too busy and having too wonderful a time to be homesick. Well, I know you wanted to boost my morale—but remember I'm *never too busy* or too active to keep thinking of HOME and all those I love. But I know it's not good to let it get the best of me. And I try never to mention it in my letters—'cause there isn't a thing any of us can do about it. But I do get lonely—and homesick, lots of times. But so must our boys over seas. And they can't go out to a movie, or for some fun dancing or have a high ball, or anything to divert their mind for awhile and I can! So we won't complain. After all I never forget why I'm here TO DO MY SHARE. . . .

All my love and kisses and prayers, Your baby, Eunice

Alameda, California

Dear Mom and Dad,                                      June 6, 1944

Well, it seems as tho THAT day—"D-Day" has arrived! Great, isn't it?! Dad, I sat practically all night by the radio in the Lounge with my ear glued to the speaker. And the best part of it was *I just knew you were listening too!* . . . I took Eisenhower's communique No. 1 in shorthand (thinking it would be longer, of course—than one sentence!) Then I heard him speak—and King of Norway, as well as Prime Minister of Belgium and the Netherlands. By 2:30 San Francisco time I went to bed. . . . I wasn't at all tired this morning. I guess I'm still excited about it. . . .

Say, I now have *another* job. I told you, I guess, that I was temporarily in the Insurance and War Bond office? Well, now I'm back in the Air Center as a stenographer. . . . Today was my first day. . . .

Love and Kisses to all, Eunice xxxxxx

Alameda, California

Dear Mom and Dad,                                  September 15, 1944

. . . Oh, I mustn't forget to tell you about how *I talked to the* CAPTAIN today! We always have "Captain's Inspection" on Fridays, but it's usually just a zone inspection made by other delegated officers, Lt. Cdrs. and down. Today, actually the Commanding Officer, Capt. Boone, came on the inspection. Also Capt. Iverson, Medical Officer, and another Capt. whom I don't know—accompanied by the Chief Boatswain, Chief Yeoman, and several Masters-at-Arms, and our new Chief Staff Officer, Comdr. Ingram. When they went thru our office, I happened to be there all alone. I jumped to attention. He scrutinized the place and then without giving an "at ease" or "carry on" command, he began to question me. "Are you in charge here?" "What is this room considered?" "How is the Guard Mail delivered?" "How often?" and numerous others. Well, I promptly answered all questions *fully* but as *briefly* as possible and with plenty of "sirs" added. Now that I think it over I should have used "Captain" every time in place of "sir"! He seemed satisfied with my answers. Later I happened to mention to the Communications Offr., Lt Cdr. Post, how nervous I was. He said he could hear the questions and answers (he's in the next room) and that I sounded perfectly poised—and my answers were excellent! I was somewhat thrilled down deep inside. You see this CO is so stern and G.I., and it's good to be that way in his presence.

Oh, yes, I must say too during my "interview" with the Capt., our new Chief Staff Officer seemed pleased with my replies, too! He's another whip-cracking G.I. guy! But that's the type I *really like* to work for! Mr. Ziglar was the Lion of Radium Chem. [Eunice's place of employment prior to her enlistment in the WAVES] until I began working for him. But because I didn't flinch or shrink in his presence he toned down a bit! However, I'll never be that close to Capt. Boone or Cdr. Ingram to even try to tone them down! . . .

Love and Kisses, Eunice

Dear Mom and Dad,
Alameda, California
January 16, 1945

Well, I came thru my blood donation 4.0! I wasn't a bit nervous, and I didn't faint either. I'll admit near the end I got feeling weak as the nurse took off the tube and needle out of my arm, took the pillow from under my head, and put a cold cloth on my forehead. But she got more than the expected 300 ccs. They are going to use my blood for plasma. Only the type "O"s were being accepted for whole blood this week, it seems, and I have type "B." Later, when we were having coffee and donuts I had to help a girl who fainted—she was from my office. I thought she'd be bringing me home, but instead I practically had to bring her home! We came home in the same bus we went in. I figured I had better not take any chances at staying ashore so I came to the barracks. But I surprised myself, I'm feeling fine; did a washing, too! . . .

Love to all, Eunice xxxxxxxx


Eunice McConnell and William McDermott, who was also in the Navy, had begun dating prior to their enlistment in the service. In fact, the couple were on a roller-skating date in Brooklyn when they first heard the news of the Pearl Harbor attack. They became engaged in December 1944 and planned to marry the following April. In early February, when William McDermott was unexpectedly transferred from his submarine base in Groton, Connecticut, to Mare Island near Vallejo, California, the couple pushed up the wedding date to February 6, 1945.

TELEGRAM—*Alameda, California [January] 31, 1945 2:55 p.m. Bill Coming West. Wedding planned Feb 6th. Have Fr. Churchill wire statement to Fr. Sheehy NAS Alameda of my baptism and freedom to marry. Wire me $100 for expenses. Wish you could be here too,*
*Love, Eunice*


After the couple returned from their honeymoon, Eunice McConnell McDermott wrote a long letter to her parents in which she provided the details of the events leading up to the marriage, the wedding ceremony, and their five-day honeymoon in Carmel, California.


                                            Alameda, California
Dear Mom and Dad,                    February 16, 1945

Now that you have waited patiently for all the news I shall try to give you all the facts and not leave out even a single detail.

First, the telegram from Bill—Oh, was I excited: But I began to make plans immediately. I asked Wanda to be my Maid of Honor and she was very thrilled. . . .

I had millions of phone calls to make—order the cake, make arrangements for the reception, and a room for Tuesday evening, and plans for our honeymoon at Carmel, and photographs, etc. I was a very busy girl.

Bill arrived on Sunday and there was quite a bit of excitement at that moment, too. . . .

We had a lovely evening together just talking a blue streak. And in the Pepsi Cola Canteen we accidently met a fellow who knew me.

He had been an elevator operator at GE! His wife had just had a baby son and he had just made a record of his voice to send to her [the Pepsi Cola Canteen was one of several places where service personnel could make "Voice Letters" to send to loved ones and friends]. . . .

Monday we had to see the Chaplain in the morning, and get our licenses in the afternoon. Tuesday morning I went over to the station dispensary to see the doctor. I thought it would be wise to find out some things and it was a good way to occupy my morning. . . .

Valerie and her Mother were in the back room of the Chapel waiting for Wanda and I to see if they could do any last minute things for us. Her mother sings professionally and asked to sing at my wedding! She sang, "I Love You Truly," and "Because." Wasn't that just wonderful? I was so thrilled! . . .

I had a very pretty book of mine decorated as a prayer book, with a white orchid on it and some other small white flowers surrounding it, and white streamers down front. It was beautiful. You'd have liked it too. . . .

Well, I don't really remember walking down the aisle, but somehow I remember the first minute I could see Bill standing down front, and everyone said a big smile came over my face. . . . When it was all over and we were standing at the back of the church I cried because you folks were not with me. I guess it looked funny for the bride to cry, and Billy was so cute when he tried to comfort me. . . .

The orchestra [at the reception] asked if they could play a piece in our honor and so I asked for "Always." They announced our name and the hour of our wedding over the microphone too. We really had a wonderful time. And I hope you liked the picture of the two of us cutting the cake. . . .

The first day at Carmel we went to the beach and walked along for about a mile and then back. Then later in the afternoon we hired bikes and went out to the Mission. . . . The second day was even more beautiful, and we went horseback riding all morning. . . . That evening we "did the town". . . .

Gosh, Bill is so wonderful to me and we are so happy. We just know our entire life will be like a honeymoon. We seem to agree on everything. . . . Everyone still tells me I was a beautiful bride, and everyone likes my ring. . . .

Love and Kisses to all, Eunice xxxxx


                                                        Alameda, California
Dear Mom and Dad,                                        April 18, 1945
. . . Did you hear President Truman speak last night? I did. Not much of a speaker, after listening to Roosevelt for 13 years, but I guess

we have to get behind him with as much push as we did Roosevelt, if we expect the same results. I'm glad I had at least one chance to vote for F.D.R. . . .

Your Baby, Eunice xxxxxxxxxxxxxx

                                                    Alameda, California
Dear Mom,                                            May 8, 1945

. . . I got up at 6 a.m. to hear Truman and Churchill proclaim the official V-E day! Great, isn't it? Well, I guess it won't be too long now 'til we get those Japs. . . .

Love again, Eunice xxxxxxxx

In June 1945, Eunice McDermott was discharged from the WAVES because she was pregnant. She returned to Brooklyn to live with her parents and await the end of the war. Bill McDermott received his discharge early in 1946, and the couple moved to Long Island, where they lived until 1966. They then moved to Charlottesville, Virginia, where they continue to live. Bill worked for Sperry Gyro for forty-three years. They have five sons, five grandchildren, and two step-grandchildren. Their travels have taken them to the Caribbean and to Alaska.

PEARL "PERLA" GULLICKSON, a school teacher from Donnelly, Minnesota, enlisted in the SPARs in August 1943 when she was twenty-five years old. Following basic training at the Coast Guard Training Station in Palm Beach, Florida, she went to Storekeeper School in Palm Beach. She was then transferred to the District Coast Guard Office of the Third Naval District Headquarters in New York City. The following letters were written to her sweetheart and future husband, Fred "Hal" Halverson, a Minnesota native who had recently enlisted in the Army.

                                                    St. Louis, Missouri
Dearest Hal,                                         August 19, 1943

Left Minneapolis at 5 P.M. yesterday. . . . [Had] a grand send-off. . . . Sleeping on the train last night was swell. Had a lower bunk. I had had to take an early train in to Mpls. on Wed. morning. The girls decided that those who needed rest most should have the two lower berths available to us. Everyone seems so nice.

We were met by Spars at the Depot this A.M. and taken to breakfast. We shall muster for chow at 1:30 in half an hour.

All twenty-five of us have to remain together at the U.S.O. until we depart at 9:50 tonight. Hope we have Pullman tickets for the rest of the journey. Should get there Saturday afternoon—we were told.

For our left arms we were given bands to wear here—words say "Service Man." These give us the privilege of using everything here. This U.S.O. Service Center is very grand—large and so many different things you can do. Will have a game of ping-pong after chow. . . .

Lovingly yours, Perla

August 20, 1943

Enroute on a troop train—and I mean troops of all kinds.

Dearest Hal,

You will have to excuse a lot of things—pencil, stationery (smudgy) and penmanship. Train stopped now so I can scribble a few lines.

This train is so sooty—wonder if I can ever get clean again. We just had breakfast—9:30—It was a typical Army mess—too heavy for me. Maybe I should get over the idea of being finicky right now. Meals yesterday in St. Louis were very good.

Soldiers have been trying to take the few seats vacant in our car, but are told to go to their own cars. . . .

Am learning the official Coast Guard marching song—Semper Paratus. Think all the girls should have it learned by tomorrow afternoon.

Don't know where I'll be mailing this. We don't stop over in any town, I guess. In Memphis, Tenn., early this morning we were unhooked from that train and are now attached to another larger troop train. Never crossed my mind before that I would ever be a member on a troop train. We are supposed to arrive in Birmingham this afternoon.

Along the way we have watched the Negroes picking cotton. Erosion is a very common sight and homes in the majority are most dilapidated. Don't see how humans can exist under such poor circumstances. . . .

Lovingly yours, Perla

[Palm Beach, Florida]

Dearest,                                    November 25 and 26, 1943

Everything happens to S.K. 7 [Perla's Storekeeper School Class]! We were just honor company two weeks ago and now we have been chosen as honor company for this week when the Admiral and Dorothy Stratton [Director of the SPARs] come. We meet and welcome them. This afternoon, we have a regimental review and again on Saturday morning there is one for the Admiral. On Tuesday night we marched over to Flagler Park, our Company was the only one to get applause from the bystanders, so the Chiefs asked what company we were. Must be that fact by which they chose us. We feel grand about the honor. Also, since we are *that,* the Admiral inspects our rooms Saturday. Always wanted to be close to an Admiral. . . .

We are in full dress uniforms these days. Keeps us busy washing and ironing white shirts. . . .

Lovingly, Perla


Visits by Admiral Russell R. Waesche, Commandant of the Coast Guard; Captain Dorothy Stratton, Director of the SPARs; and Representative Margaret Chase Smith, an early and strong proponent of women in the maritime branches of the military, form the centerpiece of the next two letters.


[Palm Beach, Florida]

Dearest Hal,                                    November 27, 1943

. . . We as Honor Company were on the ramp yesterday to give a hand salute to Admiral Waesche as he arrived in a car driven by a Spar driver. Since he is a Vice Admiral, he rated ten side boys (ten tall Spars). The Rear Admiral we welcomed this morning at 0830 rated eight side boys. Both were piped aboard.

Just came back from Regimental Review. We marched first, of course, and guess we did ourselves up proud like. Comments were that the Honor Company marched as one person, like a machine. Do we feel good! Everyone says, "May we touch you?" We are the only ones to get liberty this afternoon. Little chum and I are going out for a meal and relax. I'll go to the Depot and get some information about trains going west.

Will have to cut this short, Darling.

My love, Perla

[Palm Beach, Florida]
Dearest Hal,　　　　　　　　　　　　November 28, 1943

Take your "pics"! Enclosed please find those snapshots of a certain
Spar in uniform. You get an idea of the beautiful surroundings we
Spars have. It is a marvelous training station, there's no doubt about
it and we should consider ourselves very fortunate.

　　Our days have been so busy. We piped the Admirals off ship today—the Honor Company being envied by all because we had ring
side places as we stood in file formation on the ramp. Last night Admiral Waesche, Dorothy Stratton, and Representative Margaret Chase
Smith of Maine spoke to us. They were grand! It is so nice to see
those who are at the wheel of our organization. It is Mrs. Smith who
introduced the bill to Congress for having women in the service. [Rep-
resentative Margaret Chase Smith introduced legislation creating the
WAVES.] My, she is so charming. There were other celebrities present. All of these things coming in one week have worn us down. . . .
So nighty-night.

Lovingly, Perla


　　The following letter, written after Perla Gullickson graduated from
Storekeeper School, describes her train trip to her new duty station
at the District Coast Guard Office of the Third Naval District Headquarters in New York City. It depicts the camaraderie shared by
members of the armed forces during the wartime years, and it also
details the special respect accorded to women and men in uniform by
civilians.


New York, New York
My Darling,　　　　　　　　　　　　December 29, 1943

Arrived New York City over three hours late. Had a comfortable train
ride from Chicago but just could not sleep last night. . . . Behind me
was a serg. who has four years service and had not had a furlough in
that length of time, now gets 30 days. He was just returning from
sixteen months fighting activities in the South Pacific. He related some
of the gory experiences and had with him a Japanese gun, name stamp
outfit, paper money, articles of Japanese clothing, a piece of a Jap zero
[Japanese fighter plane], Jap prayer book carried by one of them, a
Jap radio condenser outfit, English and French coins, etc. He was
quiet in his manner of speech and tapped me on the shoulder to question me about my service. At one stop they told all service men and
women to hit the chow line at a little canteen next to a depot. I saw

that it was such a tiny place I wouldn't attempt to make it. Here this guy comes with coffee and cookies for me and another soldier brought sandwiches. Have been treated wonderfully along the way.

There were about twenty WAVES, Yeomen, coming from Oklahoma and on seven days' leave. They were a lively group and moved from one end of the train to the other. They invited me to play bridge with them but never did get a game rolling. I was glad because I was too tired to have to hold up the Coast Guard tradition of being always on the alert when I was tired. Most of them got off at Philly.

A lady got on the train at Philly and was seated "kitty-corner" from me. . . . She said she would like to take me in a cab to the place I wanted to go—of course, I didn't refuse such a kind offer and here I am. This place was/is a lovely old mansion of some millionaire who has given it to the Women of the Services. It is artistically decorated for the Christmas season and offers all the conveniences one could desire in such a place. There is a charge of fifty cents per night for a bunk. In my room are mostly Wacs and Marines, with three girls from the Canadian Wacs. They all seem pleasant and are spending furloughs in New York for over the New Year holiday. . . .

I think of you so much when I'm all by myself and in a big, big, city. I'm thrilled to pieces though and hope I'll actually be stationed in the city.

Your Perla sends all her love, P.G.

New York, New York

Dearest Hal,                                     December 31, 1943

Nearly the end of '43. . . . Bertha and I have a small room here and as next door neighbors have two nice gals that were neighbors at the Biltmore [SPARs were housed at the Biltmore Hotel in Palm Beach, Florida]. We are the only Spars on ninth deck of this hotel [Embassy Hotel]. The first four decks have been taken over by the Coast Guard. Guess our number hits around three hundred and eighty. We have liberty until 12:30 on weekdays, after work hours 8 a.m. til 5:30 approximately and on Saturday and Sunday nights it is 7 a.m. This is very liberal. We have to hit the deck at 0630 and have breakfasts here. Breakfast is served from 0700 until 0800. We eat lunches out and receive sixty cents a day for each day (work) added to our pay. Evening chow is served from 1730 til 1830. All Sunday meals can be eaten here. We even get two subway tokens for each day. Quite a well-planned set-up for us.

Miss you so much. This seems so much farther away than Palm Beach was. Could be I miss those daily letters! . . . My New Year's first wish is this—to see you again and soon.

Lovingly, Perla

New York, New York

My Darling,                                        May 1, 1944

The beginning of the fifth month in this city. The time has never passed so fast before and each day brings us closer to the end of the war. We kids talk of wearing hash marks [a sleeve insignia awarded after four years' service in the Navy and Coast Guard] but we hope that actually won't happen. Wouldn't that be a riot.

Yesterday four of us went to the Polo Grounds to see the Giants win over the Brooklyn Dodgers in a good ball game. Saw two home runs in the one game and there were some exciting moments throughout. We ate the usual "corruption" of peanuts, pop, hot dogs, etc. I limited myself to the first, for the other gets too messy. I learned how to keep a score card from the man next to me. He talked like a New Yorker—boid for bird, etc. Got a kick out of the conversation among his friends. Every once in a while one would say—"just heard my son got his bomber safely to Africa"—"just heard my son ——" These are proud *fathers* and they should be too.

There is a gate into the Polo Grounds which said that it was free to men in service. We didn't have any tickets but walked in there as cockily as you please and when the ticket takers saw us, they said, "This way for servicemen *and women*" and bowed us through the gates. . . .

Do so wish you could get to New York. We could have the time of our life and I'd love having you get lost with me as I'm sure that is what would happen if I tried to show off! Can always find the way back, though.

Time to knock off and get to work with that Monday morning feeling.

Loads of love, Perla.


On June 7, 1944, while on leave and furlough, Perla Gullickson and Hal Halverson were married in Minnesota. In December 1944, Hal Halverson was shipped to the European Theater of Operations, where he remained until April 1946. Perla Halverson remained at her duty station in New York until November 1945, when she received her discharge from the SPARs. She returned to her home in Donnelly in time to share Thanksgiving dinner with her brother, who had just returned from the Pacific. Perla and Hal Halverson were reunited in the spring of 1946. In 1948, the Halversons took over the operation of the 1871 Halverson family homestead in Cottonwood County, Minnesota. They moved to Minneapolis in 1963, where Hal Halverson worked for Midland Cooperatives, a farm supply cooperative for the upper Midwest, until his retirement in 1983. The Halversons traveled with People to People to Russia in 1979 and to China in 1981. In 1986, they spent two months in Bolivia with Volunteers in Overseas Cooperative Assistance in Third World Countries. They continue to manage the family homestead, now designated "a century farm." They have two children and three grandchildren.

MARY CUGINI, the oldest of five children, was born in Brighton, Massachusetts, just two months after her parents immigrated to the United States from Italy. She enlisted in the Women Marines in January 1945 when she was twenty. Prior to joining the Marines, she worked as a statistical clerk for the War Department in Boston. After completing boot camp at Camp Lejeune, North Carolina, she was transferred to Arlington, Virginia, where she worked as a bookkeeper and auditor with the Quartermaster Department. The following letters, written to her younger sister, Dena Cugini, exemplify the wartime esprit de corps of women and men in the Marines. They also provide a provocative account of how life in the Marines contributed to one young woman's coming of age during World War II.

Camp Lejeune, North Carolina
Hi Dena!                                              January 16, 1945

How are you these days? And how are your heart-rending romances coming along. Say listen, we just had mail call and I haven't heard from anyone. Get on the beam and hurry with that mail. . . .

Tonight our platoon has special detail on the time we are supposed to have to ourselves. Boy! Is this the works. It is one swell place. My tan is coming back again. We are outside so much. They stress the necessity of health a lot. You should see me eat. I'm a regular chow hound.

Our D.I. [drill instructor] (male), is keeping his eye on 3 girls in my Co. to see which one should lead the Co. So he's on me. It's because I had drill before. . . .

To be very truthful, I don't get lonesome, but I do miss you all! (when I stop to think—but here, the Sgt. does that, we just obey). It's some great life and I love it. I'll be happy when I graduate though. Rumor has it that we'll be sent to Phila—or Los Angeles—and you can't say till we get our orders. . . .

Special Detail—Intermission. . . .

Here is one thing I'm learning in the Marines and that is to be very patient, and to be unconcerned about things that happen. So, I guess, now I'm just like you. Swell, huh? . . .

Loads of love, Always, Mary

Camp Lejeune, North Carolina<br>January 27, 1945

Hello, Dena,

Now that I have a moment to spare—and no kidding, this is the first time I had a chance to write since last Wednesday.

Tonight I'll do my washing at the laundry, so I'll iron tomorrow after church. We went into uniform Thursday. What a difference it made on all of us. The girls look wonderful. They look swell when they are all in platoon formation. Say, what a nice looking sight and nice looking figures.

Whenever we have physical training, we do each exercise to the count of 32—and they get harder every day—the class is 45 mins. long. These exercises certainly streamline our figures. My legs have nice muscles in them now. Yesterday a girl actually passed out while exercising, and so would you, if you do what we do.

We had our shots again today, and my hand and arm is sore from it. I'm rugged, though and can take it. If you . . . would get this working out, you'd certainly be a changed girl. It's a great thing to go through and it makes you feel swell to know you can take it. . . .

We had our Captain's Inspection this morning. We stand at attention and the Major looks around the squad room. When she got to me, she looked me up and down—turned to our Sgt—and said, "She is a very neat looking girl." Along came Lt. Boling. She looked at my feet and read me off because my feet were not at a 45 degree angle. So you see, that's what the Marine Corps is—a very highly disciplined organization. That's why a Marine is so proud and full of pride—with a training like this, who wouldn't be. I could write on for hours. When I get home remind me to let you in on a few things especially on personal hygiene. They teach you everything you can think of. A doctor gives us lectures on that and they tell you all—forgetting nothing. If I tried to tell you it would sound vulgar because we use medical terms. . . .

The priest told us here in church that a chaplain's job is to keep us all clean, splendid, boys and girls, as when we left home and to marry someone who has lived your kind of life and someone from home. . . .

You know they put "salt Peda" in our food, as they do for the fellows. [Saltpeter, the common name for sodium nitrate and potassium nitrate, was supposedly used to suppress sexual activity. Rumors that saltpeter was added to the food of service personnel were ubiquitous during the wartime years.] No wonder I don't care to go out (kidding). Anyway a girl in the service deserves a lot of credit because she's doing a grand job.

These fellows who hate girls in the service—are fellows who help give the girls a bad name and they are the ones who know. . . . A uniform doesn't make you do wrong—if you weaken—it proves that you are worthless and no one, or thing, is to blame for it, but themselves. So beware of a fellow who says "I don't give a darn" because they don't give a darn about you, either. . . .

Love to all, G.I. Mary

Camp Lejeune, North Carolina
Dear Dena,                           February 10, 1945

. . . This camp is a city in itself, only it's full of military personnel. You salute, walk, minding your own business and have fun too. I'm afraid I've lost my civilian attitude. You'd probably think I'm too reserved now. We are taught to be that way, because the PUBLIC has an eye on service people. Especially gals in the service.

It makes fine men and women out of us, and no kidding, Dena, I feel like a woman of the world and I'm not afraid of anything or anyone (Rugged, huh?)

My girl friend, Ginny, says, "No wonder our hats don't fit us—our pride has gone to our heads." Ha Ha.

Well, kid, we fall out for another training film. Of course, all our films are restricted.

No more chatter for now!

So long, Mary

Camp Lejeune, North Carolina
Hello, Dena,                          February 11, 1945

. . . Last night we drilled for two hours after class. We are the "banner" platoon so now we do trick drill [a "banner" is often awarded to the platoon which makes the best appearance while marching]. It's loads of fun. When we hit the sack last night, we were all asleep in two minutes. Then, before you know it, someone yells, "Hit the Deck." This Marine Corps keeps us in trim. . . .

Being in the service like this, Dena, makes me broaden my education, and see how other people live. Honestly, it's worth it. Sometimes

I wish I didn't know so much, but they teach you all this for your own good. A guy or a girl in the service is no dope. Wish you were with me to share my experiences. . . .

Give my regards to those civilians in your office.

A Member of the "Banner Platoon." Marine Mary

In the next letter, Mary Cugini reports on her new assignment as a bookkeeper with the Quartermaster Department in Arlington, Virginia.

[Arlington, Virginia

[Dear Dena,]                                              Early March 1945]

. . . Since I've been in the service, I gained 10 lbs, grew ³/₄", and take 1 size wider in shoes. This training did me good. But I miss being outdoors. I get so sleepy in here. It isn't that I don't go to bed early enough because I usually hit the sack at lights out time, that is, if I don't go to Wash., or dancing. So far, I haven't had any real dates. Funny how I don't care to have any. Huh! Well, marines don't go in for that kind of stuff (who am I kidding). As far as I can see, we Marines are too reserved. You'd have a better time with a sailor. . . .

I like it here in Arlington though. No tall buildings. Up here, on the hill, we can see the layout of the city and the White House. There is a Stage Door Canteen across the street from it. That's another place I'm going to. . . .

[Love, Mary]

Arlington, Virginia

Hello Dena,                                               March 19, 1945

. . . Last night my girl friend Ginny from boot camp and I went to a USO in DC. I played ping-pong with a soldier and talked with him. He said I was the first Marine he ever talked to. In the meantime, Ginny met some sailors from Chicago, her home town. So we went to a huge roller skating rink. At first I felt funny because I didn't want to go with sailors—they are kids. The Marine Corps sophisticated me. But, I am glad we went, Dena, it was swell. All servicemen there. My friend Don taught me to skate and I didn't even fall once. He did. Ha! Ha! The music was surely pretty as we skated along. We wore shoe skates. There were pretty lights and couples skating. Gee, it was so much fun! Well, we ate, and then they took us to the camp. Outside our gate they were so surprised to find out we needed liberty passes and had to be back at 0100 hours. Then they saw the guard at the gates who checks us in. They know now how the Marine Corps operates. . . .

Loads of love, G.I. Mary

Arlington, Virginia
Hello Sis,                                    April 17, 1945

. . . [President Roosevelt's] funeral was so impressive. The Marine band in dress blues led the funeral. Then came the midshipmen from Annapolis, sailors, soldiers, Wacs, Lady Marines, Waves and then the President. Gee, Dena, what a sight! [We] . . . brought out with a salute when the Pres. went by. Only two sets of colors in the whole funeral. The marching was slow. The Lady Marines were real smart. Their lines were real straight. The girls here had to be over 5' 6" tall to be in the parade. . . .

Lenny called me up again Sunday. Gee, Dena! What am I going to do. I don't want to marry him. Not for a long time and he can't wait. The service life changed me a little. I have so much I want to do now—and this is my chance. I am planning to go to school evenings and take up aviation. I guess knowledge is my first love 'cause now I don't care to be married. There is plenty of time for that . . . . The service has taught me to keep my feelings to myself. So that sometimes I don't know how I feel. Anyway, time will tell. What does Ma say about Lenny? . . .

All this sounds so exciting and so on, but you know, Dena, I sure get awful lonesome. It's no easy life I assure you. All the nice times I have, I need, or else you go nuts, if you don't go out once in a while. Now I know how it is to be in the service. Boy! Do we gripe, but darn it, I still love the Marine Corps. See what I mean? It gets under your skin. . . .

Love, G.I. Mary

Arlington, Virginia
Dear Ginnie and Dena,                          July 2, 1945

Here's all the scoop to this date. Thanks for writing to me, cuzin, I guess you know that mail from home brings your morale up to 100%. Was tempted to call home long distance, today,—but I'd hate to reverse the charges. . . .

I'm glad to know the pictures came out good. What's the matter with you civilians. I'd give anything to be one now. No Kiddin'— you'll never know how G.I. Joe lives, unless you *become a G.I. Jane.*

Some life—gee! you learn so much—and ya' get *rugged!!* Do you think I did???? I know I hardened to a lot of things—no more soft heart. . . .

Loads of Love, Your G.I. Mary

[Arlington, Virginia]
Dear Sis!                                          March 12, 1946

. . . By the way Dena, I am getting too independent for my own good. Maybe it's a good thing, because now no one pushes me around and if I don't like the way people treat me, I tell them so. The Marine Corp sure has done wonders for my character. Guess I'm just as conceited as a Marine too, cause I've been told that too. . . .

Lots of love, Mary

Separation Center, Arlington, Virginia
Hi Sis!                                            May 23, 1946

We have to stand-by for inspection now. . . . Speak about flag waving, if I had a pen in my hand, I'd be a Marine again. . . .

In about 2 hours, I'll be a civilian. Just think, I'll be wearing civvies! Let's deck out in our best in New York. We both want to look sharp. . . .

That's about all for now. I'll write again. Am real anxious to see you.

Love to all, Your sister "Mary, a future civilian"

At the time of her discharge from the Women Marines in May 1946, Mary Cugini had received the rank of corporal. In 1949, she married Valentine Necko, a soldier whom she had met at a USO dance during the war years. They eventually settled in Asheville, North Carolina, where they currently live. They have two children and two grandchildren. Since recently retiring from her job as a weather information data processor, Mary Cugini Necko has organized a local chapter of the Women Marines Association. She is also a volunteer at the local Veterans Administration Hospital in Asheville. Writing almost fifty years later, Mary Cugini Necko observed, "My life changed forever when I stepped off the train at Camp Lejeune."[3]

---

3. Mary Cugini Necko to the authors, October 8, 1992.

MARION STEGEMAN OF Athens, Georgia, graduated from the University of Georgia with a degree in journalism and art in 1941. While a university student, she learned to fly in a Civilian Pilot Training Program. She joined the Women Airforce Service Pilots in March 1943 because she "wanted to be part of the action." After completing her training at Avenger Field in Sweetwater, Texas, she was stationed with the Fifth Ferrying Command at Love Field in Dallas, Texas. The letters she wrote to her mother provide strong evidence of the important ways World War II dramatically changed the lives of United States women in uniform.

Mother, Honey—
                                                      Sweetwater, Texas
                                                       April 2, 1943

Every day we have the same routine: mess, inspection, ground school, drill or calisthenics, mess, flight line, mess, then study until 10 at night. So there's not much to add from day to day except to tell you again what a wonderful training I'm getting.

    *Please,* somebody write me!

Love M.

Dearest Mother,
                                                 Sweetwater, Texas
                                               [April 16, 1943]

I thought maybe you'll like an idea of our daily program, so I'll outline it approximately for you (I'm not supposed to tell the exact schedule).

6:15—fall out of bed

7:00—fall in for mess

8:00—ground school formation (We take two subjects at a time. One hour each.)

10:20—drill or calisthenics (Sometimes sports in place of the latter) for one hour; alternate days

(We are subject to inspection any time of the day or night!)

12:00—march to mess

12:50—flight line formation. We stay on the flight line for five hours. Each flight period is an hour long, and soon we'll be getting in two or three periods a day.

6:00—Mess formation (We change from our uniform coveralls to "dress slacks" for this.)

8:00—Study hall for ground school delinquents (Hope I'll never have to go.)

10:00—Taps

A busy chicken, ain't I? In all that spare time I have I do a little housework: dusting, sweeping, mopping, making my bed. Then I have to wash out my personal items (things that can't be sent to the destructive laundry in town) and press same. Also write a few letters, study my lessons, and keep my hair and fingernails as well groomed as possible. In addition, it is necessary to fight my way into the shower (since 11 other people use it too) to keep the body beautiful sanitary. By the end of the day I'm pretty well whupped and even the board-like Army cot feels good to me.

Did you know that the Army is spending $30,000 on each of us to train us? When I think of all that dough being taken out of you poor civilian taxpayers' pockets, I feel right obligated. . . .

I love you, Marion


                                    Sweetwater, Texas

Dearest Mother,                            April 24, 1943

The gods must envy me! This is just too, *too* to be true. (By now you realize I had a good day as regards flying. Nothing is such a gauge to the spirits as how well or how poorly one has flown.) Where was I? Oh, yes, I'm far too happy. The law of compensation must be waiting to catch up with me somewhere. Oh, God, how I love it! Honestly, Mother, you haven't *lived* until you get way *up* there—all alone—just you and that big, beautiful plane humming under your control. I just sit there and sing at the top of my lungs while I'm climbing up to 4,000 feet—or however high I want to go. Of course, I'm too busy to sing while in the middle of aerobatics—but you ought to hear me let loose when I'm "clearing my area" between maneuvers. (We always clear the area first to make sure there are no planes underneath or close by—safety foist!)

The only thing that I know that's going to happen that I won't like is that they are changing my instructors some day soon. Mine is going on to the B.T's (Basic trainers—one step ahead of primary trainers) but maybe I'll get him again when I get to the B.T.'s. Hope so! I have no idea who my new instructor will be—I hope I'll like him as much as I do this one. He'll have to be pretty good and mighty nice though, to beat Mr. Wade's time. . . .

Smackers and much love to John, Janet, Joanna and you—M.

Sweetwater, Texas
Mother, Darling—                                    [May 31, 1943]

Get set! Prepare yourself! Because here comes another one of those slap-happy, nonsensical (??) ecstatic letters! OOOOOOO, Mom. I'm so happy I could die.

By now you know I've either (1) had a good day at flying or (2) passed a check ride. It just so happens that both are correct!

Honestly, mater, I was so scared when I climbed into that cockpit to take my first civilian check ride on the B.T. that I thought I'd vomit all over the controls. I had been running to the johnny for a nervous B.M. every ten minutes. One girl, seeing me dash in to the john for the fourth or fifth time (prior to my check ride) said to me, "You either are about to have a check ride or you're going on a cross country. Which is it?" It seems that it affects us gals the same way!

Anyway, I gave the check pilot a good ride and he told my instructor he might have an H.P. (hot pilot) on his hands. But since the only H.P's are dead pilots—proved by experience, he got his terminology mixed, but anyway, he meant it as a compliment. Happy day!

Then I went up for a ride with my instructor, and he told me to climb in the *back* seat which meant I was to do instrument (under the hood) flying—which is a very great compliment, since you aren't supposed to get instrument instruction until you are either qualified to solo or have already soloed! So he must think I'm ready to handle it alone as soon as I have the required eight hours, which makes me veddy, veddy happy indeed! I can hardly wait. He told me he thought I had the feel of the airplane now, and that I was cooking on the front burner! Also, he let me do a few slow rolls in the *B.T.* from the *back seat*—and he said they were *perfect*. Of course I came out from under the hood to do the slow rolls. . . .

While I was still under the hood, later, he said, "O.K. I'll take it, then you come out from under the hood when I tell you to." So he messed around awhile, then said, "All right, Come out!" So I came out from under the hood and *I was still under the hood!* He had flown right into the middle of a huge white cloud. More fun! So he flew the instruments until we were out of the cloud and could see again. . . .

So you see, your baby chile is enjoying life to the fullest. I have *everything* I want: my family loves me (and I'm sorta fond of it); I've got wonderful roommates—real *friends;* I'm doing what I love better than anything in the world; I'm so healthy and feel so good that it's revolting; and the men love me and I love the men! EEEEEEEEE, law! What a life! . . .

I love you all—M.

Sweetwater, Texas

Mother, Dear, [June 10, 1943]

. . . I went on my fourth solo cross-country ("X-C") yesterday and at last something happened to break the monotony. Each trip is around 400 miles and takes about three hours, and they really get boring. But yesterday a strong wind blew me off course and made me temporarily uncertain. I decided to head out for home anyway, but things started looking *wrong,* and the check points didn't jibe with the map. I buzzed a couple of towns but couldn't find the names of them anywhere, so I turned around and went back to a town I had just passed over that had an airport. I knew it was either Ballinger or another town north of Ballinger, both of which had army airports. The roads leading out of each town made a similar pattern and the fields were located in the same place in relation to the towns. So I entered the traffic pattern with a bunch of P.T's and landed. I could've tried to get on home, but I didn't have a heck of a lot of gas and I was bored with it all anyway, and didn't want to waste time. The Army cadets and instructors nearly fell out of their planes when they saw a *girl* taxiing by in a *B.T.!* I beamed at them all and got out at a hangar where I telephoned our squadron commander, after I had found it was Ballinger, after all.

While I was waiting for the call to go through, some Lt. came up and said, "You're from Sweetwater?" I said, yace, and he went on: "Have you seen Major McConnell?" I said, "No, he's not at Sweetwater any more." (He was our commanding officer before being transferred.) The Lieutenant grinned and said, "I know. He's the C.O. here, and I bet you knew it all along." His last remark was untrue, though I *had* heard that he was at Ballinger, but had forgotten it. He's a young, attractive bachelor, so I don't blame the Lt. for thinking I had stopped over on purpose for a visit with the Major. (It turned out one girl had preceded me there by about a half hour, having secretly planned deliberately to stop over and see him. Sh-H-H! No one knows.) Anyway, I nearly fainted from the news and trucked on over to see the Commanding Officer.

Meanwhile I had gotten Avenger Field on the phone. I forgot to tell you: when I taxied up and brought the plane to a stop, I dropped my map on the bottom of the cockpit, and I thought: Ah! I'll just pretend this happened while I was in the air, and then I will have had a good excuse to land. (They tell us always to put the airplane down if we lose our air map.) So that was my story after I realized that Ballinger was almost directly on course and if I didn't have some alibi they'd never believe I didn't have an ulterior motive in landing. So everyone but my roommates think I lost my map in the bottom of the plane and landed to retrieve it.

The Sqn. Cmdr. here (when I got him on the phone) told me I had done the right thing and to come on home if I could get clearance to take off from Ballinger. So I sacheted over to Major McConnell's office and got the most cordial greeting, I've ever had. He said, "Sure I'll clear you, but I refuse to do it for a couple of hours." So he took me to the PX for cokes and cigarettes and showed me around the post. He devoted his entire time to me, and then—and then! —flew formation home with me! I had a wonderful time and the girls were really impressed and refuse to believe it wasn't all planned. The major's parting words, before he took off for Ballinger after delivering me here safely, were: "Next time make it late in the afternoon!" . . .

Darling, I do miss you so! Please don't worry about me when you miss me, though, because, honestly, I'm so happy and having so much fun it's worth any chance in the world I could take. And, actually, I'm not doing anything dangerous at all. I fly well-kept-up planes in a country where no one will shoot me down, so I'm not really brave at all, though you may keep on thinking so if you like!

LOVE YA, M.


Sweetwater, Texas

Mother dear,                                    [Summer 1943]

. . . Mother, the most heart-breaking thing has happened. Jane . . . the older roommate—washed out yesterday. The check pilots said the B.T. was just too much for her to handle, that it was no reflection on her flying. . . . Some people just aren't capable of handling the faster, heavier ships and still they make good pilots on lighter planes. Poor Jane, though! Next to Shirley and Sandy she was my favorite, and I'll sure miss her and her inimitable sense of humor. She was loads of fun and it just about broke my heart to watch what she was going through while waiting for what she knew was to be her last ride. Of course, this doesn't hit her as hard as it would someone like the other roommates because Jane has a husband and daughter—but, as she put it, that didn't keep her "from being heartsick." Poor gal cried for two days, and I cried right with her most of the time. . . .

Love you *deely*. M.


The following letter describes Marion Stegeman's response to the deaths of two of her classmates in training at Avenger Field. All told, thirty-eight WASP died in the service of their country.

[Sweetwater, Texas
Dear Mom,                                   Late summer, 1943]

. . . Mother, this was to be a short letter, but now something has happened that I must tell you about.

You may just as well get used to hearing about these things, Mom, because so long as I'm in the flying racket they are bound to happen. Two of my classmates and their instructor were killed yesterday afternoon near Big Spring. . . .

This is no doubt another of those undetermined causes that brings about crashes, and no one will ever know what it was. Maybe it was one of those rare structural failure cases—no one knows. It seems likely, since the instructor was in the plane—Or it could have been that the girls were changing seats in mid air and one of them could've grabbed the wheel for support, thus stalling the plane. There are endless possibilities—Most of them things that *could* have been avoided, as most crashes seem to be. . . .

Don't worry about it, though, Mom, because it's very unusual for anything so mysterious to happen (especially *here*) and they're inspecting all the airplanes before we go up in them again.

As I've told you before, we *do* take chances, but they are small compared to those that thousands take every day all over the world. And we could fall down in our bathtubs at home and be killed, or get in a car and meet death. It's just not up to us to say where or when. You believe that, don't you? We'll talk about it more when I get home. . . .

I love you, Marion

[Dallas, Texas
Dearest Mother,                          November 22, 1943]

Don't you *ever* tell anybody, but guess where I'm writing this! From behind the wheel of my twin-engined Cessna. The day is so calm that all I have to do is trim the ship, and it flies straight and level by itself. I have my hand on the wheel and my feet on the rudders, though, in case we should hit a bump—And don't worry—I'm looking around every few minutes. Sounds like awfully hard work, doesn't it? I'm right on the beam, so I don't have to worry about navigating.

I won't make a practice of writing while I'm flying, but I thought it would be fun to show you how simple it really is. Of course, if my engine gave so much as a sputter, I'd die of heart failure right here. . . .

Love to you and Grannie, Marion XXX

The women who joined the WASP had anticipated that they would be granted full military status. During the early months of 1944, because of an oversupply of Army Air Forces (AAF) pilots and the need to build up American troop strength for the anticipated Allied invasion of Europe, pilots in training for the AAF were transferred to infantry divisions as prospective combat replacement officers. This situation, coupled with deep-seated prejudices against women in the military, proved to be the death knell for the WASP. In the next letter, Marion Stegeman refers to some of these problems.

Tallahassee, Florida

Mother, dear:                                              March 30, 1944

(1) General Arnold [Henry H. "Hap" Arnold, Chief of the Army Air Forces] says openly that the Army Air Forces has more than enough pilots.

(2) There are experienced instructors now being forced into the foot army—and others out of jobs.

(3) If I go into the Army, they could chain me to a typewriter for the duration plus six months, in spite of anything they might promise.

(4) I can't see myself running around saluting and kow-towing and obeying orders from [those who] . . . will really dish out the works to those of us who have been in only a year and will be mere Second Lieutenants. I can do what I'm told gracefully now only because— underneath it all—I know I don't *have* to.

*Summary:* All this adds up to a great deal of rationalization that has been taking place since I last saw my love. I want to marry him— now! Of course, though, I'd stay on the job indefinitely as a civilian, because I owe so much, but since the Army is forcing me to become a puppet or resign, I'm tempted to go my own way,—mine and Ned's.

I don't think the airplane will replace the man, do you?

It may be days, weeks, or months before it is necessary for me to decide. How's about a long letter of advice from you—and also please ask Aunt Helvig and Grannie what they think.

I love you, M.

Shortly after writing this letter, as it became clear that the Women Airforce Service Pilots would not receive full military status, Marion Stegeman resigned from the WASP and married her sweetheart, Ned Hodgson, then serving with the Marines in Fort Worth, Texas. After the war, the Hodgsons returned to Fort Worth to live. They have three children and five grandchildren. Marion Stegeman Hodgson is the author of short stories, feature articles, and several cookbooks. The Hodgsons live in Fort Worth.

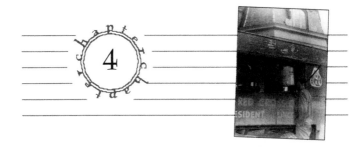

# The War Against the Axis: Italy and Germany

WHILE UNIFORMED WOMEN FROM the United States did not participate in organized combat during World War II, they were regularly assigned to postings that brought them up to or near the front lines of battle. Sixty Army nurses dressed in battle fatigues waded ashore in North Africa on November 8, 1942, the day of the North African invasion. Six weeks later, on December 22, 1942, five WAAC officers arrived in North Africa, followed by the arrival of enlisted personnel a month later. Army nurses went ashore on the Anzio beachhead in Italy on January 27, 1944, five days after the troop landings. By the time of the June 6, 1944, D-Day invasion of Normandy, over 3,000 American Red Cross workers were stationed in England, Italy, and North Africa. Four days after the D-Day invasion of Normandy, Army nurses and Red Cross women hospital workers arrived in France to set up field and evacuation hospitals. The first WACs made their way to the European continent on July 14. Two days later, on July 16, American Red Cross clubmobile workers arrived in France, preparing their first doughnuts for the troops on July 19. At the end of the war, 8,316 WACs were stationed in Europe, the largest number in any overseas theater.[1]

---

1. Robert V. Piemonte and Cindy Gurney, eds., *Highlights in the History of the Army Nurse Corps* (Washington, D.C.: U.S. Army Center of Military History, 1987), 14–18. George Korson, *At His Side: The American Red Cross Overseas in World War II* (New York: Coward-McCann, 1945), 276. Mattie E. Treadwell, *United States Army in World War II, Special Studies, The Women's Army Corps* (Washington, D.C.: Department of the Army, 1954), 360–61, 380, 387–89. Foster

Women in uniform were killed and wounded in the line of duty, captured by the enemy, taken as prisoners of war, and decorated for their bravery. The letters in this chapter depict the fortitude, stamina, and courage of the American women in uniform stationed outside the continental United States who participated in the Allied effort to bring about the defeat of fascism in Europe.

FOLLOWING THE ENTRY of the United States into World War II, the buildup of supplies and munitions in the United Kingdom grew steadily. With the tons of supplies came American military personnel to transport and organize the goods as well as to establish bases, airfields, troop training facilities, and supply depots. Military medical staffs also arrived to organize hospitals, dispensaries, and rehabilitation facilities. The American Red Cross sent representatives to serve as hospital recreational workers, to set up Red Cross clubs for American troops, and to drive clubmobiles to isolated outposts to dispense doughnuts and coffee to soldiers. Indeed, the American Red Cross workers who had come to England as early as 1940 were the forerunners of the millions of Americans who were to follow as the Allies began the immense task of overthrowing the Axis rulers in North Africa and Europe.

The first United States troops to arrive in the United Kingdom came ashore in Northern Ireland early in 1942. Army nurses assigned to this post were responsible for establishing medical facilities where both troops and local residents could be treated. In the first letter, an Army nurse assigned to duty in Northern Ireland reports on her experiences in a letter that was published in the *American Journal of Nursing.*[2]

> Somewhere in Northern Ireland
> [Late Autumn, 1942]

[Dear Editor of the *American Journal of Nursing:*]

After arriving in Northern Ireland, we lived with the British Army Sisters—nurses to you. We were called "Sister" for a bit, but we ac-

Rhea Dulles, *The American Red Cross: A History* (New York: Harper & Brothers, 1950), 425. Marjorie Lee Morgan, ed., *The Clubmobile—The ARC in the Storm* (St. Petersburg, Fla.: Hazlett, 1982), 73, 78.

2. Reprinted with permission from the *American Journal of Nursing* 43 (February 1943): 211–12. Copyright 1943 The American Journal of Nursing Company.

cepted it gladly when we realized the pride the nurses here have in that title.

Living with these girls gave us something we will never lose. Their ways and customs are only slightly different from ours in New England. And we are now confirmed tea drinkers. We are all avid knitters, which amazes the British sisters, for they thought we were rather wealthy, glamorous play-girls. . . . We have formed many fast friendships and have the greatest admiration for them and the work they are doing in all parts of the Empire. . . .

The towns are small, quaint, and picturesque. The buildings are of brick, stone, or cement, as the wood supply is limited. The people are gracious and hospitable to us. . . .

Our soldiers make good patients and I have enjoyed my work here so much. Our organization is wonderful, the officers capable. I am proud to be a member [of the Army Nurse Corps]. . . . The hospital—of Nissen huts—is quite like Army hospitals at home. There is real team work and companionship here.

R.N., Army Nurse Corps, Ireland

On January 27, 1943, the 149th WAAC Post Headquarters Company, a handpicked group of highly qualified enlisted women who had volunteered for overseas duty, reported to General Dwight D. Eisenhower's headquarters in Algiers. Described by the press as "the first American women's expeditionary force in history," the women in this unit were assigned to the Signal Corps and the Central Postal Directory.[3] The next two letters were written by members of the 149th to their former commanding officer, Captain Charlotte Morehouse, stationed in Conway, Arkansas.

                                                        [North Africa]
Dear Captain:                                           July 8, 1943

From day to day, your three WAACs have tried to get together on a letter to you, but we are still individualists, apparently, and so the time has never come when we could sit down together and tell you all about things. . . .

We are still together and we have remained the best of friends throughout the trip. Our trip over was a real pleasure cruise and quite a relief. . . . We had to sleep in our zoot suits (coveralls) on certain nights, and we couldn't go a foot without our life preservers, but otherwise I couldn't ask for a more pleasant or happy cruise. . . .

---

3. Treadwell, *Women's Army Corps*, 361.

Since we have been here we have moved our quarters several times but we hope we are settled now for a few weeks. The place we are in now is quite convenient, we have cold showers every day and plenty of room. Our last 'home' was not so convenient and we had to carry our bath water, wash in our helmets, etc. . . .

We are in one of the most modern cities of North Africa, and we have French hairdressers, French restaurants, and French and English movies to go to. . . . Our jobs are interesting, our work is confidential, and we think it is important. . . .

Thank you again for your many kindnesses. I hope I will have the pleasure of "soldiering" under you over here.

Sincerely, Frances C. Pope

North Africa
Dear Captain Morehouse:                    July 9, 1943

. . . Our crossing was excellent. . . . When we were in the danger zones, we slept in our coveralls, with full gear at the foot of the bed. Fortunately, we didn't have to use it. Full gear consisted of life preserver, gun belt with water canteen, first aid kit, steel helmet and overcoat. Some load. . . .

If you want to try something really intricate, try French plumbing or the lack of it. It absolutely defies comprehension. You're in the middle of a lovely bath, shower, or shampooing your hair, and Presto, the water goes off for the day, so there you are, all lathered and no water. I've bathed in a steel helmet, out of my canteen, etc. You have no idea what a marvelous aid to living a steel helmet can be. We bathe, drink wine, do our washing, use it for a seat, wear it for protection during air raids, carry food in it, etc., etc. With regard to the water situation, since it's so unpalatable (when you can obtain it), I can readily understand why people here drink wine the way we do water. It is the only alternative. . . . The coffee looks and tastes and is of the consistency of Karo syrup, but you grow used to it. Personally, I think that G.I. food is the best in Africa, even though it leaves a lot to be desired in the way of variety. There are two things I'll never eat again when I leave the Army, beans of any description whatsoever, and in any form, and corned beef. . . .

I still swear the mosquitoes get on their hands and knees to crawl under my netting and persecute me. It feels as if they take a hunk out of me and go sit on the foot of the cot and chew it. They collaborate with the flies in a potent and well executed technique. The mosquitoes attack, and the flies counter-attack, with my frame the battlefield. . . .

Everyone gripes fluently, with a beautiful richness and direct simplicity!!! The vocabulary of the G.I. language shames Webster for eloquence. Occasionally we deviate a little, when our control deserts us. . . .

Sincerely, and with true affection, Beatrice Page


After graduating from Kahler School of Nursing in Rochester, Minnesota, in December 1941, June Wandrey joined the Army Nurse Corps. On March 18, 1943, she arrived in Oran, Algiers, where she soon learned "that war was worse than hell." As a combat surgical nurse, she was close to the front lines of action in North Africa, Sicily, Italy, France, and Germany. Wandrey earned seven World War II battle stars for campaigns in Tunisia, Sicily, Naples-Foggia, Rome-Arno, southern France, Rhineland, and Central Europe. When the war ended in Europe in the spring of 1945, Wandrey was stationed in southern Germany, caring for the survivors of Dachau. The letters that follow appear in Wandrey's World War II memoir, *Bedpan Commando: The Story of a Combat Nurse During World War II.*[4] Her letters to her family provide graphic accounts of the harsh combat conditions experienced by Army nurses assigned to field and evacuation hospitals in the North African and Italian campaigns. (For another letter by Wandrey, see pages 225-226)


                                                    Somewhere in North Africa
Dear Family,                                        March 22, 1943
. . . My ancient French . . . comes in handy when talking with the Arabs and French natives. . . . I trade my cigarette ration for fresh fruit. Combat troops get a free pack a week. Otherwise they cost five cents a package if you can find some. The doctors are annoyed that I don't give my ration to them. You've never dreamed of such poverty and filth in all your life. I worked in the slums of Chicago and the stockyards clinic area and that was bad, but this is 100% worse. . . .
Please say an extra prayer for me tonight, A scared June.


Wandrey and her colleagues were soon transferred to Tunisia to be near the actual fighting in the North African war. They traveled in a long, arduous motor convoy of Army trucks.

---

4. June Wandrey, *Bedpan Commando: The Story of a Combat Nurse During World War II* (Elmore, Ohio: Elmore Publishing Company, 1989).

[Near Bizerte, Tunisia]
Dear Family,                                            April 25, 1943

While the Easter Parade did Fifth Avenue, we picked our creaking
bones up from the hard earth, splashed water in our faces and decked
ourselves out in fatigues and Little Abner's [coveralls]. Leaving Assi-
Ameur, we were ready to start by truck and ambulance convoy across
North Africa to the fighting front. . . .

We saw our first wire fence today. Otherwise they plant trees or
cactus for fences. Olives, figs, oranges are all along the way. . . . Today we rode for 13 hours and, believe me, the hard ground felt mighty
good when we camped for the night. We bivouacked at Afreville in
the village park, Oran Province. . . .

My neck got disjointed trying to see out of the ambulance window.
Nine nurses and their luggage in an ambulance is impossible, in-
humane. . . .

June


[Somewhere in North Africa]
Dear Ones,                                              April 27, 1943

We bivouacked at Setif, Algeria Province in a fragrant sheep field.
. . . Breakfasts and suppers were the only hot meals during the day,
unless we put our C ration cans on the motor and melted the grease.
We were allowed 30 minutes each meal to choke on the usual cold,
greasy concoctions. Beans taste more like beans every time I eat
them. . . .

Love, June


Somewhere in North Africa
Dear Ruth,                                              May 31, 1943

V-Mail is safer as it is flown over and doesn't run the risk of being
sunk.[5] I celebrated Memorial Day by going swimming at Ferryville.
The beach isn't bad, but not as nice as the one at Bizerte, which is
really sandy. . . .

Surprising how one becomes used to an air raid. It's about the

---

5. V-Mail was a special form of correspondence developed by the United States
government during World War II to help save space in scarce wartime transport.
Letters, written on specially designed 8½ by 11-inch stationery, were photo-
graphed, and the film was shipped in canisters overseas. The developed film was
then distributed to the troops as a letter in the form of a 4 by 5½-inch photograph. V-Mail made it possible to transform letters with a bulk weight of 2,500
pounds into film weighing only forty-five pounds.

most spectacular thing one could possibly see. You are so fascinated watching the anti-aircraft fire and the light from bursting bombs that you forget all about your own personal danger. No one panics. They are so well trained. There isn't any confusion, no one makes an outcry. On a pitch-dark night, an air raid looks like a big Fourth of July celebration. You can hear the enemy planes overhead, and when they get in the beam of the large searchlights, you can see the planes, if they are low enough, even the pilots. But then it is too bad for them. The enemy planes usually drop flares so that they can see where they want to bomb. It's eerie watching the flares float to earth. It's then that I wish our huge Red Crosses were as big as the Empire State Building.

By the time you get this, you will be out of school for a nice, long vacation. In a way I envy you, but just for a minute. While this war is on, I just couldn't be back in the States, despite the inconveniences we experience. They are so desperately in need of surgical nurses. . . .

Wide-eyed as always, June

Allied troops launched the Italian Campaign with amphibious landings in Sicily on July 10, 1943. Two days later, June Wandrey went ashore. What follows is one of the first letters from Sicily she wrote to her family.

Scorching Sicily

Dear Family,                                                   July 19, 1943

. . . My stockingless legs are scratched to shreds. About two hundred yards from my vantage point on the hillside, the hospital tents are going up in a race against darkness. I think we should all join a circus when we get back to the states. At the head of my cot, I have my one piece of hand luggage, on top of that are my two field packs. For a pillow, I have a parachute seat. . . .

The moon should be up soon, it has been so gorgeous the past few nights, but it gives the enemy too much help. I prefer the real dark, dark nights.

Fondly, June

Poor Sicily

Dearest Family,                                              August 14, 1943

Working like slaves. Too tired to write and it's always too dark to see when I get off duty. We were so close to the lines we could see our artillery fire and also that of the Germans. The Jerries have poor aim

today. Shells landed in front of us and behind us. I'm well and as
happy as one could be in this set up. Glad I have lots of energy. Don't
know how the older nurses stand the pace. . . .

In our pell-mell existence, we received our first naval casualties. A
ship right off shore from us was bombed and strafed. Even our den-
tists were doing minor surgery we were so swamped. We have surgical
priorities that must be operated on first: belly or chest wounds take
precedence over orthopedic surgery or some simple debridement [the
surgical excision of dead tissue and foreign matter from a wound].
Even if the patients are the enemy, if they fit the category, they come
before our soldiers. We have surgical auxiliary teams that come to our
unit to do the surgery. They work non-stop 'til the shock wards are
emptied of patients. The doctors were specialists in chest, belly or
orthopedic surgery. . . .

Working in the shock wards, giving transfusions, was a rewarding,
but sad experience. Many wounded soldiers' faces still haunt my memory. I recall one eighteen year old who had just been brought in from
the ambulance to the shock ward. I went to him immediately. He
looked up at me trustingly, sighed and asked, "How am I doing,
Nurse?" I was standing at the head of the litter. I put my hands
around his face, kissed his forehead and said, "You are doing just fine,
soldier." He smiled sweetly and said, "I was just checking up." Then
he died. Many of us shed tears in private. Otherwise, we try to be
cheerful and reassuring.

I've seen surgeons work for hours to save a young soldier's life,
but despite it they die on the operating table. Some doctors even col-
lapsed across the patient, broke down, and cried. There are many dedicated people here giving their all.

Very tired, June


On January 27, 1944, five days after troop landings on the Anzio
beachhead, Army nurses waded ashore. For the next four months, the
fighting on the Anzio beachhead was fierce, and June Wandrey was
one of many nurses who worked under heavy enemy fire at Anzio.

                                                            [Anzio
Dear Betty,                                          Mid May 1944]
Sorry for the long delay in writing, but we moved again and have been
very busy. We are restricted to the immediate hospital area as it isn't
safe to leave. There is nothing to write about but the wounded. We
live down in the ground in sand-bagged damp, smelly foxholes, like
moles in a blackout. Each hole is big enough to accommodate one

army cot. It is timbered and sandbagged on three sides and the top. It's cozy, confining and for this area I'm happy to be underground. A pyramidal tent tops it. The hospital generator is sandbagged, even the mail tent. Patients, in the past, have been killed in their cots from strafing, etc. The ward tents are also dug down into the ground, but there are no sandbags overhead. It's just not possible. The huge Red Crosses are supposed to be protection, but the enemy doesn't always observe that convention. The tent mate I had in Sicily . . . was wounded in the leg here at Anzio. . . .

Love, June

The work of Army and Navy nurses and Red Cross women assigned to the North African and Italian campaigns was often very dangerous. On September 13, 1943, a clearly identified British hospital ship, *Newfoundland,* was bombed by German planes and set afire as it was en route to Salerno. Army nurses and Red Cross workers were aboard the ship when it was hit. Before the ship sank, the women were rescued by British vessels and evacuated to Bizerte, Tunisia. The following letter was written by Peggy Telford, one of three Red Cross hospital recreation workers aboard the *Newfoundland.*[6]

Bizerte, Tunisia<br>September 15, 1943

Dear Thompson,

. . . The crosses on the ship were well-lighted and so far as we know no military ship was in our vicinity. At 4:45 AM, bombs fell near us, no hits. At 5:15 AM two more bombs were dropped, one in the sea and one upon the bridge of the *Newfoundland,* a direct hit. The bomb started a fire which, because the fire fighting equipment was totally disabled, spread quite rapidly over the forward part of the ship. About half of the life boats were put out of commission. At about 5:25 the evacuation of the ship began and all those not killed by the bomb were safely evacuated to the *St. Andrew,* another British hospital ship standing by. Miss Richards received bruises and abrasions and was suffering somewhat from shock. Miss Clement and I were totally unhurt. All three were taken off in boats along with the nurses.

The *St. Andrew* fed us, bunked us, and gave us all the supplies available, including cigarettes, clothes for the nurses and Red Cross

---

6. The letter of Peggy Telford appears in Robert H. Bremner et al., "The History of the American National Red Cross, Volume XIII, American Red Cross Services in the War Against the European Axis, Pearl Harbor to 1947" (Washington, D.C.: American National Red Cross, 1950), 244.

who had been unable to dress before abandoning ship, administered first aid and hospitalized those who had been hurt. They supplied most of the girls with soap, toothbrushes, handkerchiefs, bags similar to our ditty bags, matches, etc. The *St. Andrew* proceeded immediately to Bizerte bearing the survivors.

[Peggy Telford]


During the spring of 1944, twenty-five Navy nurses under the command of Lieutenant Clyde Pennington landed in North Africa to staff a base hospital at Oran, Algeria. Later, Lieutenant Pennington was also charged with staffing dispensaries in Bizerte, Palermo, and Naples. In the following letter, published in the March 1945 issue of the *American Journal of Nursing,* she reports on the difficult work of Navy nurses stationed in the Mediterranean.[7]

[Oran, Algeria
January 1945]

[Dear Editor of the *American Journal of Nursing:*]
I have just returned from a visit to the Navy nurses on duty in this area. The nurses in Palermo have done very good work. In ruined Bizerte they were having bubonic plague. The people [Tunisian war refugees] who are just beginning to return could not get food unless their ration cards certified each individual in the family had been vaccinated. In an open court they intently watched the needles being boiled in a two-quart boiler and carefully placed in a coffee can lid. The Arab attending this detail was dirtier than anyone you have ever seen, but his technique was good and he saw to it that his part of the work was not contaminated. I visited Palermo and then Naples and Rome, and then returned to Oran. . . .

I was quite anxious to return to my Quonset hut hospital in Oran where there were stoves for warmth, electric lights to read or work by, running hot and cold water, and plenty of blankets on the beds. . . .
Lieutenant Clyde Pennington, Nurse Corps, U.S. Navy


Helen K. McKee of San Antonio, Texas, joined the Army Nurse Corps in August 1942. After undergoing military indoctrination and combat training at Camp Forrest, Tennessee, she was assigned to the

7. Reprinted with permission from the *American Journal of Nursing* 45 (March 1945): 235. Copyright 1945 by the American Journal of Nursing Company. Page Cooper, *Navy Nurse* (New York: Whittlesey House, 1946), 129–32.

300th General Hospital Unit and arrived in Bizerte, Tunisia, on September 4, 1943, where she worked as a nurse anesthetist. She was transferred with the 300th General Hospital to Naples, Italy, in November 1943. In August 1944, she was sent to Foggia, Italy, where she worked at the Sixty-first Station Hospital until the end of the war in August 1945. (For other letters by McKee, see pages 46-49.)

Although Army and Navy nurses and Red Cross workers were not officially engaged in combat, they, on occasion, felt the need to protect themselves. Shortly after arriving in North Africa, Helen McKee wrote a letter to her father in which she described her experiences firing weapons.

[Somewhere in North Africa]<br>October 18, 1943

Dear Pop,

Yesterday I was wishing for you—had more fun. Went up to see the boys in the grove (olive) just above us and I took *my* weapon with me. The Lt. up there was a Sporting Goods salesman in civilian life and he just loves to tinker with arms. He taught me to tear mine down, clean it and put it back together, load it and shoot it. It was my first experience with a pistol, but must say, I like it. I fired an Italian gun first, and had more fun with that. The youngster that took me out to fire it, took it for granted I knew nothing about a gun, and I played the game thru until he picked a target, than I slung the gun up and took my aim, fired, hit a bulls eye. Believe me, he was flabbergasted— he had expected to see me sitting down, I believe, but after that he let me finish a round of ammunition. I am going to fire another gun, but think it not advisable to write about it. Will tell you when I can. . . .

All my love, Helen

The 300th General Hospital in Naples became a central point for the study of the use of penicillin on battlefield casualties. At the time, the "wonder drug," which was flown in from England, was not yet widely available for use in the United States. McKee wrote several letters to her parents about the great potential of this new drug.

[Naples, Italy]<br>November 19, 1943

Dearest Mom and Dad,

. . . What do you think! I gave the first anesthetic given in the 300th General Hospital on foreign soil, here in Italy. We fought hard to save this dreadfully wounded lad, even to flying in from England a supply of a new drug, which is not yet available in the U.S. It is called Penicillium. Sorry to say, however, I don't believe he is going to make it. Honestly, it tears your heart out by the roots, but we don't talk about that. The morale must be kept high. No one dares to let down. Of course, we have endless numbers of shrapnel wounds, flak wounds, bullet wounds, and every sort of wound imaginable, but we also have the same old things, hernias, tonsils and the regular old line from Camp Forrest, but most of it is war casualties. Working over here is completely different from stateside. We have learned that each little scrap must be utilized. We have learned to work without materials we considered absolutely necessary. For instance, we have no CO2 for our machines, no soda-lime, no pins, either straight or safety pins, and paper is so scarce. . . .

Love from the depths of my heart, Helen


Dear Folks,                                            [Naples,] Italy
                                                   February 8, 1944

. . . Last Sunday I had one of "those" anesthetics— He had a piece of flak in his lung, and Col. Pistine chose Sunday morning to do it. As a matter of fact the anesthetic began at 0845 hours and ended at 1145 hours. It was—as all Lung decortication we do, Pentothal induction, Gas Oxygen and Ether Anesthetic, endotracheal, and it went swell. Honestly, I say as I tap on wood, —I've not had a gummed up anesthetic in the past six months. . . . These cases are very difficult, naturally, since the good lung is on the down-side—the bad lung on the up-side— You can see how position in this case would tend to create a difficulty in respiration. There are so many complications that can arise out of this type of operation—really is quite a risk, and I usually feel a little unstrung just before I start one, but once I get a gas mask in my hand there is something just short of a miracle—I get just as calm and matter of fact at that point and from there on I never have a qualm over my anesthetic. . . .

I love all of you very much—Helen


Darlings:                                              [Naples,] Italy
                                                     March 7, 1944

. . . Within the past three days our department has slackened its work to some degree. We are still definitely earning our room and board plus the salt in our bread which is obviously lacking at the present, and will be until a ship comes thru. . . . We do, thank goodness, still have salt for the plaster casts, and after all that is the most urgent use we have for salt here. We are still doing a land-office job on orthopedics and the Penicillium ward is going strong. The results are marvelous—most unlimited results. Can you imagine these grossly infected wounds, covering a large area, mostly with bone fragments protruding up into the wounds. We used Penicillium on some of these cases for a period of three days and found the wounds practically clean and healthy at the end of the third day period. At soon as the bone fragments are covered over, we do skin grafts. It looks so crazy to see these great depressed areas. This is so new we have not been able to tabulate our case outcome, but we are watching with eagle eyes. Day before yesterday I spent my entire day with Penicillium patients. So far as we know the drug is not obtainable for civilians either in the States or in England. Of course you know that Penicillium is of English origin. I am sure you have read of Prof. Fleming's discovery, quite by accident [British bacteriologist Alexander Fleming developed the antibiotic penicillin]. One thing on which a great deal of stress is laid is the fact that Penicillium without proper debridement of wounds just does not work. To sum the matter amounts to this: early and sufficient debridement, Penicillium 3 to 5 days—skin graft! . . .

Mom, by the way I did want to tell you, while on the subject of skin grafts, these grafts are secured either by a plain everyday little basting stitch or a tailor's stitch off a spool of #40 white cotton, just the same cotton you have there on the machine in the brown wicker basket. You still have it, I hope. . . .

Love, Helen


My Dear Ones,                           Same Place [Naples, Italy]
                                        June 25, 1944

Been rather neglectful once more. . . .

Mauldin's cartoons are now being syndicated— Keep an eye open for my favorite character, G.I. Joe. [Bill Mauldin was a famous World War II cartoonist who depicted bedraggled G.I.s, Willie and Joe, in his cartoons.] One of the best ones depicts services held in a slit trench. The shells are flying and bursting overhead and close by—you can see the oncoming shell—to quote the Padre, "Forever and ever, Amen, Hit the Dirt." Now these characters do look like our own boys. As we have seen them both times we have gone on Evac. status and at the time I was with the Evac. . . . Both during the Anzio beachhead landing, and Cassino drive, our patients came to us with only first aid care. We debrided their wounds, did the amputations, took slugs out of the livers and removed their shell fragments from all vital organs. We cleaned them up—killed the maggots in their wounds and *put them to bed between clean sheets.* Four days later they came back to surgery and we did our Gen. Hosp. job. Secondary closure, bone plating, grafting both bone, and *much, much* skin. We operated on their spines removing foreign bodies and fracture fragments (Laminectomy). We decompressed head fractures and dug foreign bodies from the lungs, diaphragm, every type of surgery imaginable.

Let me tell you about one evening last week when I was on call and was sent for. I had a very acutely ill patient on the table. — Terribly irregular pulse and his breath was so labored. I put him to sleep with Pentothal and in short order, Major Burford had the man's heart in his hands and I do mean literally— He opened the outer lining of the heart (pericardium) removed a shell fragment the size of the last phalanx of my little finger, picked out numerous scraps of woolen O.D. [olive drab] shirt and drained the abscess. The patient's pulse was regular on leaving the table and his blood pressure had picked up wonderfully— That, my dear, was pericardiotomy. One of the few I've ever seen in my life. . . .

Love, Helen


Helen McKee was transferred to the Sixty-first Station Hospital at the Army Air Forces base in Foggia, Italy, in late August 1944. The following letter contains a description of her airplane ride to Foggia, as well as information about her new duty station.


                                                        [Foggia,] Italy
Dearest Mom and Dad,                            August 28, 1944

On Friday Miss C. called—my orders had come thru, and Saturday morning, I went down to the Mediterranean Air Transport Force and ran thru the long line of screening, booking, priority and what have you, returning with passage on an Army Transport Plane, scheduled to leave Paglioni Field at 2:45 the following day, Sunday, and on that day, having gathered an arm load of pay vouchers, orders, credentials, etc. fulfilled my engagement at the field. . . . A British officer adjusted my safety belt. We all lost a bucket of body fluid thru perspiration. The great metal doors finally closed with a bang, the motors were roaring now and a little later, we were taxiing across the old plowed up field. We turned. The engine went into full swing and there the ground beneath began to sink, or we to rise. We climbed steadily until we were quite high indeed, and that out of necessity since we came over or thru a mountain range. Once here flew thru a pass, with great ragged cliffs on both sides of us. A lovely sight, and so on, our uneventful journey, until much too soon, journey's end. . . .

This morning I went to breakfast with the Chief Nurse and then thru the formality of papers, etc., and then to surgery. I found surgery a pleasant little place. One gas machine, and only two tables compared to our eight at 300th. I am the only anesthetist but there are M.D.s to relieve me. We don't have my time worked out yet. We get a day off a week. We don't start to work, on casualties until the planes come in from their missions during the afternoon. That is what I am waiting for now—the flak happy boys. . . .

Much later. Bed time. *I've just finished unpacking.* Just think of it. I have a room to myself! . . .

Love, Helen

With the celebration of V-E Day on May 8, 1945, Helen McKee's thoughts turned to her homecoming. Yet, as this letter demonstrates, the horrors of war would remain with her for many years to come.

[Foggia, Italy]

Dear Folks,                                                        May 23, 1945

. . . Now, my dears, just *when* we will leave Italy or whence from here, only God and the European Gen. knows. The best policy is to start your own rumor, that, at least, pleases you, and you have the fun of stringing yourself along. . . . If we go out to the Pacific we will have our 30 or 45 day TD [temporary duty] in the States, but, now I am of the opinion we won't be going to Tokyo because we have to close the house over here. . . .

May 25

Put this letter aside to keep my date with Lane. This war lasted just a bit too long for the boy—just another few days and he would have had his eyes and his left arm. I shan't forget, very soon, the sudden nausea that overtook me when I realized it was a matter of enucleating both eyes instead of one. Of course, we are supposed to be accustomed to seeing handless arms, and your ears should be deaf to the groans of agony from these poor souls, *but alas the two years of combat has not hardened the heart,* nor hardened the soul but, praise God, their days of "wounds due to enemy action" have ceased. You know at times I still feel sort of numb to anything except the fact that casualties have ceased to be. I don't think anyone felt closer, or shared the pain of these boys than we, the A.N.C. Now we have finished our jobs, we've seen the war thru. We are tired and ready to come home. . . .

Time to go off duty now. All my love to each and every one of you. My dears, may it not be too long before we stand, hand in hand—my family—
Helen


Helen McKee returned to her home in San Antonio at the end of the war and married T. W. Giles in 1954. She worked as a nurse anesthetist until her retirement. She now devotes much of her time to helping medical students understand the needs of geriatric patients. Helen McKee Giles was recently honored as the number-one volunteer at the University of Texas Health Science Center in San Antonio.

DURING THE EARLY YEARS of World War II, Margaret Anderson, a 1936 graduate of Texas Tech in Lubbock, Texas, worked as a civilian dietitian at an Army hospital, the William Beaumont General Hospital, in El Paso. In 1943, civilian dietitians and physical therapists who worked in Army hospitals, who could pass the physical examination, and who had no children under fourteen years of age were "requested" to join the Army. Qualified women were granted blanket commissions and then assigned to duty with the Medical Department of the Army. At the end of the war, 1,623 dietitians were serving in the Army.[8] Margaret Anderson was one of the dietitians "requested" to join the Army. With virtually no military indoctrination or training, she was assigned to the 300th General Hospital in Naples, Italy, where she served from January 1944 until March 1945. She was then transferred to the Sixteenth Evacuation Hospital near Florence, Italy, where she remained until the end of the war. The letters that follow were written to her mother in Post, Texas.


Somewhere in Italy
My Dearest Mother,                                    January 31, 1944
. . . Did my 2 boxes and 3 bags arrive from El Paso? and the sweater? Have you notified Post Dispatch of my change of address? I will be

---

8. United States Department of Commerce, Bureau of the Census, *Statistical Abstract of the United States, 1946* (Washington, D.C.: Government Printing Office, June 1946), 221.

so happy when my mail starts coming. . . . I have almost finished another piece of needlepoint. I expect I would have had it finished but I don't have any more thread with me. . . .

I shall repeat that I had a most enjoyable trip to the very end. This hospital is beautiful and immense. I haven't been out of the building since I got here. There is no need to. We have everything all in one building. I have a room with 4 other girls and we have no extra room. In fact I don't know what I'm going to do when I do get my belongings, but I'll manage.

Margaret


[Naples, Italy]
Dearest Mother,                                             April 28, 1944
. . . I guess Daddy saw in the paper about the eruption of Vesuvius. It was an awe inspiring sight. We were just lucky to be over here to see it. . . . The 300th is the Vanderbilt Unit and is quite a famous unit. We are set up in one of the finest and most beautiful hospitals in Europe. It was completed in 1939 at the cost of $16,000,000—an Italian T.B. Sanitorium. As a matter of fact, the Italians still have about half of the building. It is huge. . . . We are planning to go to the opera "Aida" Friday. . . .

Love, Margaret


[Naples, Italy]
Dearest Mother,                                             May 23, 1944
. . . You said you were glad I was strong enough to resist the temptation of smoking. You are a bit wrong. I don't smoke just because *I don't want to.* There's no temptation unless you want to. You know I don't say G____ and D____ either for I don't like to hear it but sometimes I get mad enough and I have to "resist" that temptation. Also I don't get drunk just because *I don't want to.* I see so much of it and it is revolting to me. I can't see any use in it. A few drinks are ok for those who can stop but too many people can't stop—or so it appears. To say the least, they don't.

Here again most everyone smokes. Offhand I can't think of anyone who doesn't occasionally at least. I know a few doctors here who don't but I think of only one girl and I think she does on occasion. Few women smoke like they really enjoy it. Most look as if they are doing it for effect. For others it's a nervous habit. . . .

Love, Margaret

[Naples, Italy]
Dearest Mother,                                    November 25, 1944

. . . We had a 3 star general as a patient the other day. I was going
up to see him when I met our C.O. He asked if I were going up to
see the Gen. and took me in to introduce me. He said now I'll
tell you this about Miss Anderson— We have 4 messes—patients,
officer patients, enlisted and officers. Each morning she goes around
to see which has hot cakes and gets a stack for her breakfast each
morning.

2 or 3 days later the Gen. was up and about and inspected the
hospital. When he came to the diet kitchen he came right up to me
and spoke. I asked if they had been feeding him alright and he said,
"Yes, did you get your hotcakes this morning?" My boys got a big
kick out of it and said I was really getting up in the world when 3
stars come around to talk to me. Really I've been able to meet several
of the big ones. . . .

Love, Margaret


[Naples, Italy]
Dearest Mother,                                     December 1, 1944

Guess what? Pogey's [a young Army man whom Margaret met in December 1943 while at Camp Patrick Henry, Virginia, awaiting ship-
ment overseas] been here. Isn't that wonderful? He made a most wonderful record. Received a D.S.C. [Distinguished Service Cross], a
Legion of Merit, and was decorated with a medal of the same class by
the British. Of course, I've always thought he was wonderful but now
I know he's "my hero." He gave me a wedding ring he took from a
dead German, as a souvenir. . . . I remembered he was in the Pacific
but I had forgotten where. It was at Guadalcanal. Now he wants to
go back to the Pacific again. But I don't want him to—needless to say.
I won't say "no" though.

When he called I recognized his voice and you know I knew him
only Dec 16–22, almost a year ago when I was at Camp Patrick Henry.
I wish you could see him. I think you'd like him. He calls me Peg or
Peggy. . . . We are not making any plans. He says it's too early for
that. He said it would be alright if he was killed but he might come
back with a leg or so off. He was 33 on the 22nd of Nov. I can't think
of anything else at present to tell you about him. I guess a page and
one half is enough. . . .

Love, Margaret

[Naples, Italy]

Dearest Mother,            January 13, 1945

. . . Guess who I got a letter from since Christmas—Arthur Lee Henry. You know he has never answered a letter I've written since the one when I broke our engagement back in June 1939 and I have written to him many many times—always sent Christmas cards, etc. . . . I'll probably never love a man like I loved him, but some how I just couldn't marry him now regardless if he did want me. He was perfect. I guess I just wasn't good enough for him.

Pogey is so different and yet so like Arthur Lee. He fairly worships me yet he has a mind of his own. I think we need each other to be complete. We each have our weaknesses and can strengthen each other. He doesn't know just what he wants to do after the war. He was working in a bank until it went broke during the depression and didn't work at anything very long at a time afterwards. He's been in the Army about seven or eight years now. He's put aside several thousand dollars which he will have to start something. I asked where we were going to live and he asked where I wanted to live. I told him he was going to make the living and it was for him to decide. He has no ties anywhere and wants to live in Texas, he thinks. . . . Pogey is from Chicago, Ill. . . .

It won't be long now until I'll be 29, then in another year I'll be 30—an old gray-haired woman. I can just see people whispering, My! Doesn't Margaret look old? I am getting gray-haired, there's no question about that. But I remember you had gray hair rather young. I really don't mind, in fact I like it. People used to guess I was much older than I was—now I'm guessed years younger. . . .

I love you, Margaret


[Naples, Italy]

Dearest Mother,            February 13, 1945

. . . I really have news for you. I am so enthusiastic about the change. I am getting a transfer from the 300th. I don't know just when my orders will come through, so I can't give you my new address yet. I am going north and will be able to see the rest of Italy. I'll be in the Florence area, don't know just exactly the spot.

I think I would be bored stiff to stay here as long as I know this unit will be here. I will be with an Evac. unit. They will be the first to leave the theatre and then I can most likely get some assignment in the Pacific like I want.

This unit has been dissatisfied with their dietitian and wanted someone who could stand on her own feet. My feet are big so they thought I could do the job. She was not a good mixer. Apparently I am so—they asked me *if I wanted the place.* It was just as sudden as getting my overseas orders. At first I didn't know if I wanted to move or not, for I like these people, but the more I thought about it, the more I wanted it. Of course, there will be some disadvantages—probably some I don't know about, but I've had nothing but the best so far so I can certainly stand some hardships.

I will be the only one and that has its advantages. It will be different in every way. . . .

Love, Margaret


                                                        [Northern Italy]
Dearest Mother,                                          March 2, 1945

. . . I am very happy with the move. This is a good unit and the people have been wonderful to me. . . .

The first day I made the menus, a number of people came to me and said I saw the menu and I know you made it because it's so different. I really felt good. I can't see what the other dietician did. They said they never saw her except when she came to pick up the report. . . . She got up very early and worked split hours. I was afraid I would be expected to do the same. I was tickled pink when I found that they didn't want that—in fact, frowned against it. I go to work at 8. I will make a lot of changes but I must do them gradually so that they won't notice them so much. . . . I am the only dietician and what I say goes. I have a good bunch of boys to work with. Everyone is very nice and cooperative. This is a very good evac. set-up. Everything is very convenient and comfortable. . . .

You should see me in my new combat clothes. I'll have to send you a picture as soon as I get one. We wear trousers and laced boots and pile lined jacket. I hope I can keep both of them. They are quite something. I bought my own boots so those will be mine. They are "dillys." They are natural leather therefore practically white which marks me as a rookie. That isn't really true, though. They can't see why I wanted to leave the set up I had but I had reasons which I have told you already. I want to see what's going on here and when it's over, go to the Pacific. . . .

Love, Margaret


                                                        [Northern Italy]
Dearest Mother,                                          April 22, 1945

. . . We've moved up north. The Germans are really on the run. They are farther from us now than they were before we moved. We are in a beautiful spot—a clover field with a row of grapevines on one side of our tents. I'm sure we won't be here long. I'd just as soon move on. I got only glimpse of Bologna, but I don't care to go back as all Italian cities seem to be alike as far as I'm concerned.

It rained night before last and a bit last night too so it's a bit juicy out today but the sun is shining beautifully and it will soon be dried up.

We have 2 nurses from the 300th on detached service with us. One is a rather good friend of mine. I was happy to see her and she to see me. She had been moving every 3 or 4 days for 2 weeks, and was getting a bit fed up with it all. She is in the same tent with Janey, Bantie and me. . . . .

Love, Margaret


                                                            [Northern Italy]
Dearest Mother,                                             May 12, 1945

. . . Congratulate your only daughter. She has reached the *top*. I've been promoted to "first" lieutenant. I've been knowing for sometime but wanted to wait until I got it before letting you know. It pays to be where your work is appreciated. . . .

Love, Margaret


Even at the battle fronts, women and men in uniform found the opportunity to fall in love. Sometimes these romantic liaisons led to marriage, but, on other occasions, romances were broken off. Many "Dear John" and "Dear Jane" letters were written, but few of these missives were saved. Margaret Anderson was the recipient of a "Dear Jane," and she provided an exact transcription of the letter for her mother. What follows is a rare example of a "Dear Jane."


                                                            [Northern Italy]
Dearest Mother,                                             May 27, 1945

. . . Well, I didn't get a letter from you but I did get one from Pogey. I will write it exactly as I received it—

"I've tried to write this letter a dozen times but didn't know quite how. I guess the best way is to be frank and tho I know it will hurt, I must tell you.

When I returned to the States, I became engaged to a girl I used to go with. Being with you made me forget her but when I came back and saw her again, I knew.

I'm sorry, Peggy, and I can't tell you in words how much of a heel I feel.

I hope we can still be friends and I'll never, never forget you."

It was written from China on May 15. Yes, it hurt for I had counted a great deal on knowing him better for I felt a very strange something drawing me to him. I have always had a strange feeling that we might not *love* each other if we knew each other better tho. Even though I don't have the incentive to go on as I did, I'll survive. . . .

Well, I guess that's all for now. Leaving in the morning for Venice for 6 days. The leave came just at the right time to help me forget. . . .

Love, Margaret

[Northern Italy]
Dearest Mother,                                                   June 6, 1945

. . . I'm not just spoofing about this rank business. There is 1 Major—Head Dietician in Washington—1 Captain in Europe. Not all the General Hospitals in the States have Captains, only those with training schools. So—I have gotten just as high as is possible until there is Congressional legislation providing for us. Any capt. and majors you may have seen in the paper were likely nurses or WACs. I sometimes wish I had gotten into the WACs instead, but one would never know. For sure, it burns me up to see nurses that I know personally that are as high as Lt. Col. who have no more personality and social grace than a post. I've known them as Lts. . . .

I didn't have a very good time in Venice, except buying things. Hence the chief nurse will help me get to France if possible, I think. We don't know if we will leave here next week or 6 months from now. Time only will tell and personally I don't care what happens. . . .

Love, Margaret

[Northern Italy]
My dearest Mother,                                            August 15, 1945

I was awakened this morning with the news that the war is over. We've been expecting this for days now. At last, I guess it must be a reality which makes everyone very happy. . . .

It shouldn't be too many months before I'm home. . . . The occupational troops are all round us so we don't lack for entertainment. All in all I'm very happy. If I can just get home by Christmas.

Love, Margaret

# Part II

# Overseas Assignments
## and the End of the War

WACs stationed in remote outposts played an important role in speeding up mail delivery during World War II. This photograph depicts WACs sorting the mail at the Central Postal Directory in North Africa. 1943. *(Women's Army Corps Museum)*

Major Charity E. Adams inspects members of the 6888th Central Postal Directory in Birmingham, England. The 6888 was the only African-American WAC unit to serve overseas during World War II. March 1945. *(Women's Army Corps Museum)*

Army nurses assigned to a field hospital arrive in France on August 12, 1944. *(National Archives)*.

Army nurses of the 110th Evacuation Hospital take a bath in a European stream. 1944. *(Army Nurse Corps Historian's Office, U.S. Center of Military History)*

*(Bottom left)* Four days after the D-Day invasion of June 6, 1944, Army nurses arrived in Normandy. This June 14, 1944, photograph shows Lt. Stasia Pejko making a last-minute check of blood bound for France. *(Army Nurse Corps Historian's Office, U.S. Center of Military History)*

Lt. Patricia Basinger, a surgical nurse with the Twenty-seventh Evacuation Hospital in France, writes letters home in her tent. November 1944. *(Army Nurse Corps Historian's Office, U.S. Center of Military History and Armed Forces Institute of Pathology)*

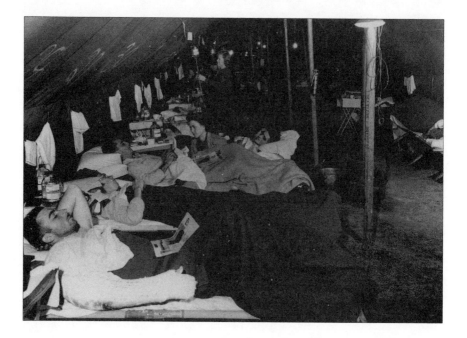

An ANC nurse at work at a field hospital shock tent. Somewhere in Europe. 1944[?]. *(Army Nurse Corps Historian's Office, U.S. Center of Military History)*

*(Bottom left)* A post-operative ward of the Twenty-seventh Evacuation Hospital. Somewhere in France. November 1944. *(Army Nurse Corps Historian's Office, U.S. Center of Military History)*

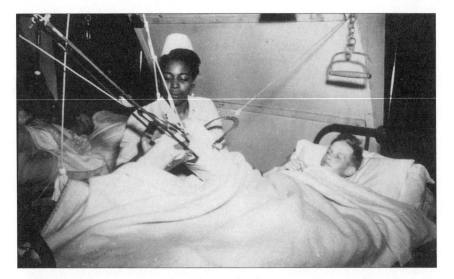

African-American nurses stationed in England were assigned to care for German prisoners of war. As one black ANC nurse noted in a letter to Mabel Staupers, Executive Secretary of the National Association of Colored Graduate Nurses, this was a " 'bitter pill' to have to swallow." In the photograph, Lt. Florie E. Gant tends to a German prisoner of war at the 168th Station Hospital in England. October 1944. *(National Archives)*

Red Cross Clubmobile Group A about to embark from Southampton, England, to Utah Beach. Clubmobile Group A went ashore on July 18, 1944, two days after the first group of Red Cross clubmobilers arrived in Normandy. *(Dwight D. Eisenhower Library and Oscar Rexford)*

Red Cross workers distribute coffee and doughnuts from the clubmobile "President Lincoln—Abe." Somewhere in Europe. 1944. *(Dwight D. Eisenhower Library and Oscar Rexford)*

---

African-American nurses with the 268th Station Hospital undergoing training in Australia before being assigned to advanced bases in the Southwest Pacific. February 1944. *(National Archives)*

In April 1944, these African-American ANC nurses with the 268th Station Hospital were sent to Milne Bay, New Guinea, where they remained until May 1945, when they were sent to Manila, Philippines. *(Courtesy of Prudence Burns Burrell)*

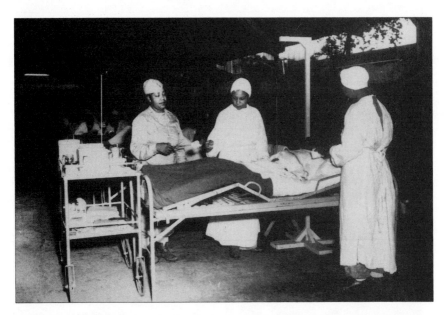

Surgical ward treatment at the 268th Station Hospital, Base A, Milne Bay, New Guinea. This was the only all-black hospital in the U.S. Army. Lt. Prudence Burns, ward nurse; Lt. Elcena Townscent, chief surgical nurse; and an unidentified nurse attend Sgt. Lawrence McKreever. June 1944. *(National Archives)*

Two of the 252 African-American women who served with the overseas division of the American Red Cross during World War II. The women are entertaining G.I.s in Assam, India. August 1944. *(National Archives)*

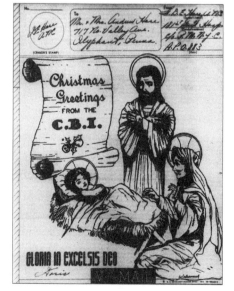

A specially designed 1944 Christmas V-Mail from the China-Burma-India Theater sent by Lt. Doris E. Hare, ANC, to her parents in Olyphant, Pennsylvania. *(Authors' collection)*

First group of women ARC workers to be assigned to Kunming, China. The women flew from Calcutta, India, over "The Hump" to Kunming. January 1944. *(Special Collections, Texas Woman's University, Denton)*

ARC worker Rita Pilkey and two of her colleagues help G.I.s decorate Easter eggs at the Red Cross Club in Kunming, China. 1944. *(Special Collections, Texas Woman's University, Denton)*

The ARC Canvas Cover Club, established by Rita Pilkey and her colleagues, at Luliang, China. 1945. *(Special Collections, Texas Woman's University, Denton)*

Five WACs, among the first to arrive in Manila after its liberation, inspect the damage to the famous Pier #7. March 14, 1945. *(Women's Army Corps Museum)*

An unidentified WAC sits beside an air raid shelter in the Philippines.
1945. *(Women's Army Corps Museum)*

Not until early 1945 were WAVES, SPARs, and Women Marines allowed to serve outside the continental United States. Two SPARs, wearing their newly issued parkas, report to the district office in Ketchikan, Alaska, to relieve shore-based officers for combat duty. *(Historian's Office, U.S. Coast Guard)*

*(Bottom left)* During the fighting at Iwo Jima and Okinawa early in 1945, the Navy began to use hospital-equipped evacuation aircraft to move critically wounded patients to large hospitals at Guam, Pearl Harbor, and in the continental United States. In this photograph a Navy flight nurse feeds Marine patients aboard an evacuation transport plane enroute from the Okinawa battlefield to a hospital in the Marianas Islands. *(U.S. Naval Historical Center)*

PFC Lorraine Turnbull, pictured here, was one of just a thousand Women Marines who, during the early months of 1945, were selected for overseas duty in Hawaii. She was assigned to the Marine Corps Air Station at Ewa, where her skills as a flight mechanic were put to good use. Ewa, Oahu. May 1945.

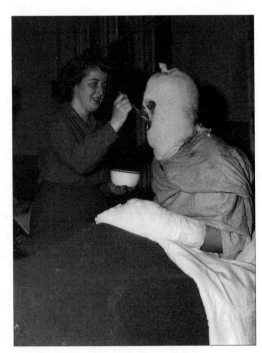

Following the end of the war in Europe in May 1945, ANC nurses were deployed throughout Europe to care for wounded soldiers and begin the work of rehabilitation. Within five hours after Army nurses arrived in Berlin on July 5, 1945, they began caring for the wounded and ill. Still wearing her field uniform, Lt. Muriel H. Stern feeds a burn patient. *(Army Nurse Corps Historian's Office, U.S. Center of Military History)*

Navy nurses held as Japanese prisoners of war from their capture in the Philippines in May 1942 until their release from the Los Banos Internment Camp in February 1945 talk with Admiral Thomas C. Kinkaid. March 1945. *(U.S. Naval Historical Center)*

Navy nurses aboard the *U.S.S. Sanctuary* care for liberated prisoners of war at Wakayama, Japan. September 1945. *(U.S. Naval Historical Center)*

With the end of the war, the return of millions of Americans from their overseas stations to the United States proved to be an immense task. WAVE June Kintzel, a flight orderly stationed at the Naval Air Station in Olathe, Kansas, often accompanied sick and wounded troops recently evacuated from overseas on cross-country airplane flights to hospitals in the United States. 1945. *(Courtesy June Kintzel Rugg)*

Marine Corporal Pat Hommel boards the *U.S.S. Consolation* for the trip from Pearl Harbor to San Francisco and home. December 1945. *(National Archives)*

In November 1945, after serving twenty-two months overseas, Margaret Anderson returned to the United States and received her discharge. After the war, she worked as a dietitian on both the East and West Coasts of the United States. She has traveled extensively, visiting all fifty states and nine countries. Margaret Anderson Piatt lives in Hood River, Oregon.

WITH THE FALL OF ROME on June 4, 1944, all of North Africa, Sicily, and southern Italy were secured from the Axis blight. The rest of Europe remained occupied, however. From airfields in England, Corsica, and Italy, bombers rained down bombs on German strongholds on the European continent. American troops stationed in the Middle East were engaged in moving vast amounts of lend-lease goods through the Persian Gulf to Russia. Work in areas as remote from actual combat as the Caribbean, South America, and Iceland contributed to the Axis downfall. United States women in uniform could be found at all of these locations, making important contributions to the winning of the war.

Florence Holtman of Indianapolis, Indiana, worked as a Red Cross club director on Corsica from August 1944 until February 1945. During her assignment on Corsica, many Army Air Forces bombers headed for European targets were based on the island. In the early spring of 1945, Holtman was transferred to Italy, where she continued to work until November. In the following letter to her parents, she describes the work of the Red Cross in helping to keep up the morale of airmen stationed on Corsica.

Corsica
Dearest Family:                                                October 4, 1944

. . . This island is poetically beautiful, great wonderful mountains surrounded by sandy beaches and an unbelievable blue sea. I have seen nearly all of it now and I don't believe I could ever tire of being here. Of course, one has to be a nature lover to feel that way about such a place as this. Five minutes away from our bivouac area, one can be in a gorge or valley with the mountains rising sheer above thousands of feet and the solitude stuns you. Fish are so abundant in the deep clear pools, Pop would be able to pick out the one he wanted before catching it. . . .

Three of us run a fine dough-nut business here, and our work is to meet the missions with hot coffee, doughnuts and a cheerful word or smile. The only realization a war goes on is when the planes have had a rough time and some do not come back, then in a day or two you miss a bright-eyed boy or two but nothing is ever said about it. Three or four evenings a week, we take entertainment or accordion and guitar players to mess tents and all sit around for hours and sing favorite songs. A lot of these men have been overseas and away from families more than two years and anything we do means a lot to them. All in all there is a great satisfaction in this work and we are much appreciated. We also arrange a lot of dances and bring native girls to them, but the whole family usually comes and we have to serve them food. . . .

Love, Florence

Late in 1943, the American Red Cross began to operate snack bars in large recreational halls for United States troops stationed in Iran. By the end of 1944, fifteen snack bars had been established. In a memorandum of January 31, 1944, Florence Joan Carey, a Red Cross worker in Iran, described the activities of one of the early Iranian Red Cross snack bars.[9]

[Somewhere in Iran]
January 31, 1944

. . . Between three and four hundred men are served each evening. The Snack Bar is in a corner at one end of a long narrow room with about twelve large tables crowded around it and three ping pong tables at the end of the room. . . . One man gets all additional refreshments for his whole table while the rest jealously guard his chair. They could eat a lot more donuts than we can supply so we try to ration them without hurting feelings.

The entertainment rises spontaneously from the crowd. One evening we had a boy playing the harmonica so some of the others went back to the barracks for theirs, and we had a concert. Another evening a guitar player sang hill-billy songs in which most of the men joined. But the greatest success has been a phonograph which Special Services installed with new dance records contributed by men who received

9. Excerpts from the January 31, 1944, memorandum of Florence Joan Carey appear in Bremner, "American Red Cross Services in the War Against the European Axis, Pearl Harbor to 1947," 140–41.

them from home for Christmas. The first night the phonograph was on, the swing hadn't been playing long when a couple of the men started to jitterbug, in another minute I was pulled out on the floor amid howls of delight and then the two Polish waitresses were persuaded to dance too. A couple of the G.I.'s took over the serving and we had a party the rest of the night. . . .

Everyone pitches in and helps—when we run short of cups, the special service officer, Lt. Molland, who is usually in the kitchen keeping an eye on things, washes dishes with the rest of the volunteers. . . . One of the men told me he never would have believed he could have so much fun in Persia—and strictly without vodka. . . .

Florence Joan Carey

Uniformed women made many different types of contributions to the winning of the Second World War. During the early years of the war, Tamaranth D. Knigin, an entomologist from the University of Minnesota, undertook important entomological research on mosquitos and tropical diseases at the U.S. Engineer Hospital in Trinidad, British West Indies, where she worked as a civilian in uniform. Because so many American troops were stationed in tropical zones, entomological research of this type was of particular importance. In the following letter, Tamaranth Knigin reports on her work to a former professor at the University of Minnesota.[10]

[Trinidad, British West Indies]

Dear Dr. Riley,                                                June 23, 1942

I am sorry for the long delay in answering your letter. I am afraid it was a combination of circumstances that caused it. Your letter took an exceptionally long time on its way and I was sent out of the colony on an entomological mission for a few months. . . .

I am sending, via air mail, a vial with *in situ* specimens of engorged Chigoes [chiggers], and some specimens of *Anopheles* [mosquitoes]. . . . Because our supply of insect pins is running low, I shall not be able to mount the mosquitoes much as I would prefer to send them that way. Let me know if you need any slides of microfilaria, we can give you those in any number you wish. I have not come across anyone whom I can recommend to you as a responsible collector. I shall try,

---

10. The letters of Tamaranth Knigin are located in the Tamaranth Knigin Papers, University Division of Entomology And Economic Zoology Papers, Folder 289, University Archives, University of Minnesota, Minneapolis, Minnesota.

however, to send you specimens of parasitological interest as often as I can gather them in reasonable numbers. . . .

[We] have worked out a new small mosquito cage for use in collecting and feeding mosquitoes. We have written a short note on it which we have to send by way of the District Health Officer to the Surgeon General for his approval. From then it will be sent to you. Would you see if you can get it published for us in one of the parasitological journals . . . ? We are sending you a sample cage. . . .

Since I last wrote to you, censorship restrictions have become more stringent. We have a new address and may no longer mention our location. . . .

Sincerely, Tamaranth D. Knigin

Flash—We have found a new species of *Anopheles* here.

The great buildup of supplies and personnel in the United Kingdom continued in preparation for the anticipated cross-channel attack into France. By the time of the June 6, 1944, D-Day invasion of Normandy, more than 1.5 million American troops were in England. To help provide for these service personnel, the American Red Cross opened 400 service clubs throughout Great Britain, operated 150 clubmobiles, and, as D-Day drew near, established "Donut Dugouts" with snack bars and informal recreational facilities.[11]

Betty B. John of Cleveland, Ohio, served with the American Red Cross in England from October 1942 until March 1944. (For other letters by John, see pages 49–50.) Her duties included helping establish new clubs for American troops. She wrote long letters to her husband, Colonel Henry J. John, a World War I veteran who served at Stateside locations with the Army Medical Corps during the Second World War. Her letters are filled with the details of her Red Cross work in England during the buildup to the Normandy invasion.

[Manchester, England]
December 7, 1942

Henry Dearest:                                      Historic Anniversary

. . . I am beginning to enjoy my work here. It is so vitally needed, for there is nothing here for the men. . . . My plans are going forward. They meet with approval and cooperation on Mr. Hanna's part. He's really good to work with. I've been holding the fort alone for a week

---

11. Dulles, *American Red Cross,* 427.

while he's at Headquarters getting us more trained staff. The two of us just can't do it alone. It's twice as big as Rainbow Corner [a famous Red Cross club in London] and has tremendous dormitories besides. . . .

Lovingly, Betty

 

 

                                                 [Manchester, England]

Henry dearest:                                          January 29, 1943

. . . Spent last night at the Allens' little doll house in Thelwell. They've been very kind to me and it's been so nice having . . . [them] to turn to for information, help or advice. . . . They seem to know just when I am most tired for their invitations have come at the two most welcome moments of my stay here. Both times to stay at night were when I hadn't had more than three hours sleep for weeks each night and had been eating the city air instead of breathing it. A night in the country, deep sleep, no noises and keen fresh air, breakfast in bed all picked me up each time. . . . But I do get very tired, sometimes so much so that I feel as tho' I'm going to fold up. This is a very noisy place—a great open well with vaulting stone steps (no carpeting) run diagonally up one side my bed room. All night troops trudge, hop, skip or stagger up and down, bang doors in the dormitory above, rats gnaw in the floor and walls because of the kitchens, (the stench was so great that the floor boards had to be taken and a cart load of ___ removed. Two workmen fainted on the job). The blackout becomes very oppressive where it's on so long. So, one just doesn't get a proper rest. . . .

All my love, Betty

 

 

                                         [Somewhere in England]

Henry Dearest:                                         March 1943

. . . The result of the RAF's [Royal Air Force] visit is that the Colonel is going to have a chance to ride in a B-26 sometime. He's lucky. I've always wanted to—even go on a mission to know what it is like—but I'm barred so far, as a civilian from even getting inside one. . . . They do a marvelous job flying at night or escorting our bombers by day, and my hat's off to them. In fact, if they hadn't been so fine and strong where they were, I'd hate to see New York and some of our coastal towns to-day, and now they're pounding Hell out of Germany. So are we, *but* and we forget this, we've never gotten to Berlin yet, and every clear night, when we hear waves and waves of Halifaxes and Lancasters [British bombers] going over, we know where they are going and the gruelling trip they have ahead of them. If they can completely demolish the heart of the enemy, they'll have no heart for fighting. So far, our effort is not so great in comparison, and we lose sight of that quite often. We have so much yet to do. . . .

I was all set to request leave for home—or resign—because it is awful to be stuck in such a tiny town forever . . . but I've been begged to stay at it by everyone— Am apparently doing a bigger job than a good many others—in fact, still the only place for combat officers whose morale *is* important to the Army—so I'm sticking. . . .

All my love, Betty


                                                    [Somewhere in England]
Henry dear:                                              April 1, 1943

. . . The sky is full. Goodness it's marvelous the might we've got here. But the flower of our men are going down one by one, and we feel it here tremendously. The good die young. I almost wish I could too.

Love, Betty


                                                    [South Coast of England]
Henry Darling:                                          May 30, 1943

So much has happened in one week that I don't know where to begin. A diary would be too long and that's about the only way I can convey to you how full my time has been.

I've been sent to the South Coast. . . . Went over to the Officer's Club (Red Cross) to write some letters. Had just started mine to you when some R.C. girls and three officers inveigled me into walking along the Promenade before lunch, insisting that I get some last minute sunshine. Well, with not more than two minutes warning, Jerry was upon us. We threw ourselves flat into an embankment at the roadside as he passed over (about 30 planes) spreading havoc as he went. I counted 20 planes and then ducked for this pale blue uniform was getting too much of a target. I did watch after they'd passed over, the bombing of the town—horrible. Then back over us again and with their machine guns, then out to sea. I spent from then until near midnight helping with the casualties, a grim sight. We converted the Officer's Club into a hospital, filled every bed with casualties, then set up cots in the ball room, then the lounges. As there was no where else to stay (I had to report to HQ on Monday morning) I left, at midnight. Two days later they were still getting bodies out of buildings. For twelve hours I had scrubbed and cut blood caked clothes off people, and made beds and started all over. That is war, and I guess I can take it. . . .

That's up to the minute news. Tomorrow brings all the headaches that go with such jobs. Getting staff, vetting staff, marketing our ration points and licenses. But there's a beautiful set up, so here goes.

Love, Betty

                                                    [South Coast of England]
Henry Dearest:                                              June 13, 1943

Getting a new club open is no picnic, and suddenly my correspondence has piled up into a mountain because I am too tired at night, and no time in the day to answer it. . . .

Other than the long hours, 9 a.m. to 12 midnight, without a break, I love this place and this work. This is the heart of excitement right now and many a time one's emotions get somewhat wrenched. Sometimes there's not much left of me by midnight after working all day at forms and more forms, petty detail, plans, organization of work, etc., having heard planes go out before starting and then three hours later hearing them come back, counting them, wondering which one of the officers you'd been speaking to the night before was one of those not present. Then serving supper and snacks as gaily as possible to glum looking men, who've lost their best friends that day and can't say so, or brought back their planes with crews half shot up.

Some of them want humor, kindness, friendly conversation, even love. Their needs are so varied, and it's hard to deal with them all, certainly impossible to solve all their problems.

Most are just youngsters. They have to be, to do what they do and stand what they take. We all seem to be waiting on the brink of something this week. Something cataclysmic and stupendous. I wish you were over here. . . .

All my love, Betty

                                                    [South Coast of England]
Henry Dearest:                                              July 18, 1943

. . . You're right. I am very close to action, and the more the Allies do to the Axis, the greater the retaliation, mostly against civilian populations, as they can't come in strong enough numbers to be effective anymore against vital war targets. For me, this is a thrilling spot, for the great long windows of the club are like an airport, where one watches planes come and go. The feelings have been described better than I can—how we watch and count them go out, knowing most of the crews by name now, and hold our breath when hours later they begin coming back. Each day greater and greater armadas sail forth.

The steady drone of their powerful engines thrills one, gives one the sense of great confidence. When you read the paper, you get all the details which we do not, so that you know our confidence is justified. If only we can increase our materiel in ever greater and greater crescendo, they'll crack, for not even they could have anything like this. The boys come into the club at night, clean shaven, calm, clear eyed, sometimes gay from a drink or two. They say nary a word. You'd think they'd been at a desk all day back in Cleveland. Comes 9 o'clock news, we turn to the radio, "The Americans have been over _____ today, So many hundred planes. The target is in ruins. The enemy gave considerable fight, only two planes lost, about 50 enemy aircraft shot down!" They never even look at each other or wink. But they're the ones who did that!! One plays the recordings his wife made on records and sent him. He's a fine pianist. Another plays records he brought from the states, The Ink-Spots are his favorites. Others bring nurses from the near-by station hospital, and chatter through the whole news without hearing it. They don't have to, they know. Still others want to chat or flirt. They're all hungry—prodigious appetites. But the appetite is not for food alone. Food is the only appeasement they can get so they eat and eat, but they're hungry for home, their families, American girls, their own girl, a normal life again. They can't be condemned for such natural desires. I call them all "my problem children" right down to the quiet, charming colonel who is their C.O. and one of their greatest pilots. . . .

All my love goes with this, Betty

In spring of 1944, Betty John accepted a commission as a captain in the Army, and she spent the remaining months of the war as a war correspondent on the European continent. The letters she wrote while a war correspondent were not saved. She is the author of several books, including *Libby* (1989), an account of a nineteenth-century Alaskan pioneer. She has also been a contributor to *Military History.* Betty B. John lives in Albuquerque, New Mexico.

WORLD WAR II often resulted in painful family separations, many of which involved daughters and sons leaving home for war-related duties. In a rather unusual reversal of circumstances, Red Cross worker Hester J. Leavitt, the mother of two college-age daughters, served in England, while her daughters attended Binghamton College in Elmira,

New York. Her letters to her daughters describe her experiences as an assistant club director with the ARC in England.

[England]

Angel:                                              January 15, 1943

I didn't get my birthday cable off to you on time because I am on a detached service job for a couple of weeks and am not on my own field. I'm on an Army post to pinch hit for the Club Director who is ill and on sick leave for a few weeks. It is quite different from the Air Corps and it is the largest club in Field Service. It has all the appearance of Grand Central Terminal. I shall be glad to get back to my own little club. Last night it was raining and I was on my way back to this club, driving the car, when I got caught in the middle of retreat. So, I had to stop the car and stand beside it at attention in the rain with no hat on, until the flags had been lowered (I guess it is called "striking the colors"). An Army post is nice though. This club closes at 9:45 p.m. so every night after I am in bed, I hear them sound taps and it sounds so *secure* and *safe*. I love it.

There is something very thrilling about watching retreat also. Makes a lump come in your throat and you are awfully glad you are an American. . . .

I have my first red stripes on my left sleeve to signify 1 year's service. Looks nice. I also have 3 little new girls to train—fresh out of Washington. They are minding the cold. We have little gas grates in each of our bedrooms which we keep burning all day and all night. We have a little separate house not far from the club and it has a living room, bedroom, and kitchen on the first floor, and 3 bedrooms and a bath on the second. It's real cute and fairly warm. We have a woman who makes the beds and keeps it clean and we eat our meals at the officers' mess. I don't eat breakfast but have a nice dinner at noon. I don't go to dinner but have a sandwich and coffee at the club at night instead.

[Mom]

The "typical" World War II American Red Cross worker in Europe was the "doughnut girl," dispensing coffee and doughnuts to the troops just behind the battle lines. The next selection of letters was written by a Red Cross "doughnut girl," Mary "Chichi" Metcalfe of St. Louis, Missouri.

Mary Metcalfe volunteered for overseas duty with the American Red Cross in August 1943. After completing her training in Washington, D.C., she was sent to the United Kingdom, where, in January 1944, she was assigned to Clubmobile Group A, based in Scotland and England. Metcalfe was among the second group of Red Cross clubmobilers to arrive in France following the D-Day invasion. She and her Clubmobile A colleagues arrived at Utah Beach on July 18, 1944, just two days after the first group of Red Cross clubmobilers went ashore. Clubmobile A then followed the American troops through Europe until the end of the war. Mary Metcalfe Rexford's husband, Oscar W. Rexford, has written an account of his wife's club-mobile life, entitled *Battlestars and Doughnuts* (1989). Excerpts from Mary Metcalfe's letters to her mother follow.[12] [For other letters by Metcalfe, see pages 223–224.]

[At Sea
Late December 1943]

Dearest Mommie:

Well, here we are finally en route. We have been on our way for five days now and so far I have not missed a meal. The first few days I felt not so good at times, but I have managed to keep eating so you don't get seasick. This is a bit difficult to do as we only have two meals a day. I eat at the second sitting which is at 9 o'clock and at 7 o'clock.

You should see the set up we have here on the ship. There are 24 of us in a cabin about the size of my room. We sleep in double-decker bunks. I have the top bunk in a corner right in a port hole. I keep a blanket draped over the port hole to keep the draft off of me. The girl who sleeps under me now is Isabelle Messinger. She is very attractive. Her husband was in the Air Corps and was killed in June. I don't know the details. She is such a nice girl. . . .

We were some fancy looking group when we were ready to go. All of us in full regalia. We wore our overcoats, steel helmets, pistol belts on which hang our canteen and first aid kit (we carried our mess kit on our musette bags) and our gas masks we wear on the opposite side from our musette bags. . . .

You will be interested to know that I have gone into Clubmobiles now. I have no idea where I shall be from time to time and when I find out I cannot tell. Censorship is very strict both on my letters and on any coming to me. If my letters don't seem to tell you as much as

---

12. The letters of Mary Metcalfe are located in the Mary Metcalfe Rexford Letters, Papers of World War II Participants and Contemporaries, Dwight D. Eisenhower Library, Abilene, Kansas.

you would like it is because we must be so careful not to say anything we shouldn't.

You will be glad to hear that in Clubmobile we wear a visor hat just like the one to my striped suit with the slacks and jacket. The color is RAF blue, a very pretty color. . . .

Lots of love to you and everyone.

[Mary]


                                                                   [Scotland]
Dearest Mommie,                                            January 27, 1944

. . . I have only been out with the Clubmobile for two days so I don't know too much yet. We leave at 8 o'clock in the morning, drive out a couple of miles to the field, park next to a hangar and get the Club-mobile all hitched up. . . . On our side there are two counters which are made up by pushing up and out two kinds of trap doors. . . . The mess kitchen at the field sends over two or three urns of coffee, what-ever we ask for each morning. We have a push thing like for chocolate sauce in a drug store where we put our canned milk to serve with the coffee. Then at the other counter we have a tray which has life savers, gum, cigarettes, and some K ration chocolate in it. Also books like the Pocketbook editions, and postcards like the ones I sent home. At the end and on the other side behind the driver's seat is a victrola, and records just above it. There is a microphone which carries the music out of the Clubmobile and also which the boys can have fun with. Next to this is the doughnut machine which turns out 7 doughnuts a minute. We make about 3 mixes in a morning which is quite a few doughnuts. Next to this is the sink where we wash all the cups. Just across from the sink is a big container for water. This is heated by a primus stove. Just under the counters are the drawers where the coffee cups are kept. . . .

There is one thing about living with the A.R.C. and that is that you have to be prepared to set off for anywhere at any time. It seems to be the policy in Clubmobile work that you only stay in one place for 3 or 4 months. . . .

Yesterday after we made doughnuts and served all morning we went to two other places and just served and played the victrola in the afternoon. This is good-bye to pretty finger nails as this is real manual labor, up to your elbows in dough, washing cups and cups and cups and fixing and playing around with all sorts of interchangeable things. I know it is going to be fun when I catch on to the way to do every-thing. . . .

Chichi

                                                    [Scotland]
Dearest Mommie,                          February 27, 1944
. . . While we were away this week we had ice cream three nights in
a row. They have a freezer at the hotel where we stayed. So we had
vanilla, chocolate and strawberry. . . .

   We get the greatest kick out of the expressions of the people. Instead of everything being wonderful as we would say, they say it's
smashing (broad A). Of course, it's aye for yes, ta for thanks, and the
best of all is when Mrs. Andrews wakes me up, she says, "Come Away
now, Mary, it's 7:45" . . . .
Chichi


   In preparation for the Normandy invasion, Mary Metcalfe and
other Red Cross workers were taught how to drive 2½-ton GMC
trucks to be used as clubmobiles on the European continent. The following two letters describe this training. Driving and maintaining
heavy vehicles was just one of the many new and unconventional jobs
assumed by wartime women in uniform.


                                              [London, England]
Dearest Mommie,                                  May 7, 1944
We arrived this morning having had little sleep in our third class
sleeper the night before. We went and deposited our luggage and then
started off for our driving. Spent the rest of the day learning to drive
those big 2½ ton GMC trucks. They are fairly easy to handle but it is
difficult learning to double clutch which you must do with all these
babies. . . .

   Yesterday was again spent driving the truck all around, in the
morning, in the thick of town traffic and out on the highway, and tank
course in the afternoon.

   I came home thoroughly worn out and took a bath and rested for
awhile. . . . Today in the driving course I drove in town this morning
and this afternoon we had a lesson in mechanics, finding out all about
the motor and what everything does. . . .

Lots and Lots of love to all, Chichi


                                              [London, England]
Dearest Mommie:                                 May 13, 1944
This has been one mad, dashing week, thus I have not had a chance
to write so much as a line to you until now. . . .

Well, the driving week ended today. I have seldom been so dusty and dirty as every day this past week after driving a GMC truck here, there and everywhere. My battle dress which I have just had cleaned is again vile dirty and my hair is so dirty I doubt if it will ever be really clean again.

Aside from learning to drive these trucks we had one afternoon of being under the hood or bonnet as it is called over here, learning all about the generator, the carburetor, distributor, air cleaner, fan belt, fuel lines etc. Another afternoon we spent lying very comfortably under the truck learning about the transmission, transfer case, differentials, pillar box . . . and a thousand other things. There were about 15 of us taking the course and we each had a GI teaching us. . . . The very first day we were out on the tank course, bumping around and going up and coming down the steepest roughest hills you ever saw. You have to use the low range gear on these hills and it is amazing to see what those trucks can do on hills and rough terrain. . . .

On Thursday we spent the whole day on the tank course and a dirtier, more tired lot of girls you have never seen than we were when we came home at the end of the day. . . .

Good night and lots and lots of love, Chi-chi

The last weeks of May and the first week of June 1944 were filled with tension. The servicemen who had trained so arduously for the invasion were at a fever pitch of expectancy. Red Cross workers and the Army nurses who would soon follow the troops to the Continent were equally tense. The huge armada being readied in southern England with the goal of liberating Western Europe was the largest ever assembled in history. By end of the day, June 6, 1944, more than 150,000 Allied troops were ashore in France. This day marked the beginning of the end of the Nazi regime.

Gysella Simon of Cleveland, Ohio, worked as a Red Cross club director in England and described her "little part" in the events of June 6, 1944, in two letters to her parents. (For other letters by Simon, see pages 256–257.)

[Somewhere in England]

Dear Folks,                                                          June 4, 1944

. . . We are anxiously awaiting the invasion, just as you are. We have worked in close contact with the Combat Forces who will make the first assault when it comes off. . . . Pray for them . . . they will need it. I shiver when I think of the boys who won't come back. I don't know whether or not I told you about a group of Rangers who were stationed here with us . . . nice clean-cut kids who have been trained for hand-to-hand fighting when it comes to that. Well, they, too, will be there. If you could see all the elaborate preparations necessary, you would not wonder about the delay 'cause there are so many little things to think about. Red Cross, too, will be well represented and while in London last week, I met plenty of girls who have been learning to drive Army trucks on which will be mounted screens for showing movies in the combat outfits. No stone is being left unturned to give them the very best of Red Cross services as soon as possible after the beaches are taken. These gals are all American girls, like myself, who have come over to do a good job, and believe me, they are doing just that. . . .

All my love, Gys.

<br>

[Somewhere in England]

Dear Folks: June 8, 1944

Well, by now you have received news of the invasion and I suppose you were wondering what little part I played in it. Can't tell you much except that we all worked like the devil. . . . Since I was located near a Navy base, we helped out temporarily and I can tell you that it was a port of embarkation for combat troops who were destined to take a part in one of the things that will go down in history. We saw their eager faces and anxious eyes and we stayed practically all night seeing that they got that good old Red Cross service right to the last minute.

We cried, of course, and had no shame for our tears. There were times when I could scarcely see for the tears in my eyes. It wasn't easy, but I did my best to hold back my feelings—a hard thing to do at a time like that. The port was filled with ships of all descriptions and to see the boys going aboard, grim and determined, was a sight which will live with me for the rest of my time. In the dead of night they sailed away and now the world knows the story. I have since talked to some of the sailors who manned the landing barges . . . they come back and forth to make their trips, and this is what they said of it. It was the neatest, biggest, most co-ordinated, slickest job they had ever seen, and the British who took part and have returned, are slapping the Americans on the back for this good job done. . . .

All my love, Gys.

<br>

For the next six months, Gysella Simon continued to work in England, where she experienced the ravages of Germany's new V-1 "buzz bombs" and V-2 missiles. Writing to her family on September 1, she remarked, "We know only too well what they are and see them every day. The agonizing moment comes when the motor cuts out and it starts to glide down. You never know where it will hit and all of us will be damn glad when our boys in France clean up the Bomb sites. . . . Since 7 AM this morning we've had 8 alerts. Everywhere you see people walking about with heads and arms bandaged—these are the lucky ones who only got hurt—god knows how many were killed."

In December, Gysella Simon was sent to Belgium and stationed with the 366th Fighter Group near the front lines of battle. In the following letter, written on December 27, 1944, she provides a first-hand account of the Battle of the Bulge.


                                        [Somewhere in Belgium]
Dear Folks—                              December 27, 1944

. . . For a resume of my own existence here it is—uncensored. Your own little Gizzie is now enjoying the distinction (and it is a great honor) of being attached to not only the *very best* but also the most advanced fighter group in Europe—this means, of course, that we are in a hot spot—especially now with the Jerry offensive in Belgium now in full swing. For the past 3 days we have been visited nightly by Jerry, strafed and bombed—always with some casualties, but your little Gizzie is still here dodging them. Incidently, the alert just sounded and I am writing in total darkness with illumination by flashlight only. Last night was really rough—you could almost hear the Jerry pilots breathe as they dive-bombed and strafed! Whew! . . . We are anticipating another rugged evening—3 alerts since I started to write this letter! Damn it, it's a nice clear night out—and that "ain't" good. . . .

    Please continue to pray for us all and to write those lovely letters. . . .

All my love, Gizzie


Gysella Simon spent a total of twenty-two months in Europe. Following her return to the United States in the autumn of 1945, she enrolled at Western Reserve University, where she received the B.A. and M.A. degrees. She taught school for eighteen years, retiring in 1982. She lives in St. Augustine Beach, Florida.

MARTHA GELLHORN, a well-known American journalist, was a passenger on one of the first British hospital ships to cross the English Channel during the Normandy invasion. Her dispatches told of her awe at being "in the midst of the armada of the invasion." She wrote about the difficulty of transporting wounded soldiers by stretcher from the shore into water ambulances and finally onto hospital ships. On D-Day plus one, Gellhorn stowed away on a water ambulance and waded ashore at Easy Red beach where she talked with the troops and comforted the wounded. She was probably the first American woman ashore on the Normandy beachhead.[13]

On D-Day plus four, the first Army nurses and Red Cross women hospital workers arrived in Normandy, where they helped establish field and evacuation hospitals. The next letter, written by Army nurse Ruth Hess to her friends and colleagues at the Louisville, Kentucky, General Hospital, describes coming ashore at Normandy during the latter part of June 1944, and life at the front lines of battle.[14] Although she expresses great pride in her work as a combat nurse, she, like many other women in uniform, also longed for the time when she could marry and begin raising a family.

[Somewhere in France]
August 8, 1944

Dear Miss Losby and the rest of you at Louisville General Hospital:

. . . I'm sitting outside my tent on the ground after a *hot* shower and a shampoo. . . . I'm just too wide awake to crawl into that bed roll under that mosquito netting to go to sleep. This is my first chance for a shower and gee, did it feel good—these cold water helmet baths are refreshing—but not very cleansing. For six weeks . . . [we] have been in France—and it's certainly been up to and exceeding our expectations. . . .

For an entire week we waited for our chance to embark—even got

13. Martha Gellhorn, *The Face of War* (New York: Atlantic Monthly Press, 1988), 109–20.

14. The letter of Ruth Hess is located in the Louisville General Hospital School of Nursing Records, Historical Collections, Kornhauser Health Sciences Library, University of Louisville, Louisville, Kentucky.

so far as watching our men board the ship—and we were left behind—
"dry runs"—galore. But finally we managed to get all set on a Liberty
ship . . . [and] for us the journey was *very* smooth. We were off the
coast of France before dinner and what a sight. Literally hundreds of
ships and capped with a barrage balloon, small landing craft and airplanes surrounded us. There was a high cliff in the distance on the
top of which you could see the black and white striped planes of ours
taking off. We embarked by way of a small landing craft with our
pants rolled up—wading into the beach a short distance. . . .

We marched up those high cliffs and about a mile and a half under
full packs, hot as "blue blazes"—til finally a jeep here, a duck [A
dukw was an amphibious vehicle] there, etc. picked us up and took
us to our area. . . .

Next afternoon we joined our unit in an enormous convoy and
started toward the place we were supposed to set up our hospital. . . .
We were eating cold C-rations. We arrived there late in the evening
and spent all nite getting ready to receive patients. Next day we
worked until 3:00 p.m. and started nite duty, 12 hours at 7:30 p.m.
For nine days we never stopped. 880 patients operated; small debridement of gun shot and shrapnel wounds, numerous amputations, fractures galore, perforated guts, livers, spleens, kidneys, lungs, —etc, everything imaginable. We cared for almost 1500 patients in those nine
days. Many of them formerly treated by the Germans while they were
prisoners. We also had numerous Germans and French.

After those nine days we reorganized, rested and parked—for two
weeks—then we moved—set up and—no patients—last Tuesday a
week ago we started taking patients again and again I was on nite
duty. Continually we operated for 5 days—597 cases and were ordered
to this area—we still had 250 patients—but were evacuated while we
packed and moved out.

We left in early evening—travelled until about midnite—when we
found ourselves ¼ mile from the front line—fireworks galore, it's
funny how you're not scared, even with all you see and hear—but
you're not—its just like watching a mammoth Fourth of July celebration if someone doesn't get hurt.

We had to turn this long convoy around and come further south.
It seems our area was still occupied by the Germans. We arrived [cen-
sored] early in the morning and set up our hospital tents again. By
noon we were receiving patients—this time our O.R. nurses are scrubbing, shampooing heads, etc. [censored].

We sent our surgical teams and our men to a German hospital that
had been captured to operate and evacuate the patients to us. All the
work they did was dirty surgery—every one was covered with pus,
gangrene, bed sores, and filth—absolutely skin and bones. . . .

Everyone of us worked, bathing 850 patients—usually with the aid of a scrub brush, changing literally thousands of dressings, reoperating and casting—making them as comfortable as possible. . . . It's really been an experience. It's lovely all day—but at nite—those d—d German planes make rounds and tuck us all into a fox hole—ack ack in the field right beside us, machine guns all around—whiz—there goes a bullet— It really doesn't spare you—you're too busy—but these patients need a rest from that sort of stuff. . . .

The life of an army nurse in combat—I love it—but just the same I'm anxious to get back home, marry my Ang. and have that family of six children and enjoy the respect of a community as a doctor's wife. . . .

We've seen all the wrecks and ruins of Cherbourg—to Rennes-Volonne, Mountebourg, Coutances, St. Lo, Caen, Charentan, Isigny, Ste. Mere D'eglise, Pont L'lobbe, Avorandes—etc, etc, etc. Every city has blocks lying in a mass of ruins—caused by our own artillery and bombing, but our Engineers are busy cleaning up—as soon as possible. . . .

Sincerely, Ruth Hess

The following letter, written by an Army dietitian, Lieutenant Mary Eaton, describes life in war-torn France during the summer of 1944.[15]

[Dear Friends,]
                                                        [Somewhere in France]
                                                        August 15, 1944

. . . It's simply marvelous to be over here in France at last after all those dreary weeks of literally waiting and training and waiting. We began to wonder if we really would get here in the end.

We landed on the beaches just as everybody else had done. It was strange to think that we were walking up that beach and over the sand hills behind it without a moment's thought of any possible danger, when just a few weeks before men had died in hundreds as they fought their way through all the cruel obstacles and heavy fire. It is amazing to see what has been done by the engineers since D-Day, and to see how quickly an area can be cleared up after the fighting is over, roads and bridges repaired, telegraph wires installed, mines cleared, of course, and supplies brought before you can turn around—almost.

We are living under canvas, of course, and have moved several times since arrival; we are now permanently located for a few months

15. The letter of Mary Eaton is located in the Army Nurse Corps Archives, United States Army Center of Military History, Washington, D.C.

anyhow, we hope, and are watching our tent hospital appearing in the fields across the road. . . .

The 1944 Normandy glamour girl is brown as a berry, wears no make-up except lipstick, and shows only trousers and shirts, leggings, and boots in her summer wardrobe this year. Headgear this year is a tin helmet, worn at all times. . . .

When we first arrived and spent our first night in the field, we wondered if we would ever come through alive. Noise, noise, noise, overhead and all around! However, it's amazing how blase one gets. We're in quite a safe spot now anyhow, and almost all the noise is ours—day and night it rolls on—supplies, supplies, supplies, troops, ammunition and overhead the constant roar of planes. . . .

Well, I've talked a lot of nonsense. There really *is* a war on over here and we're all too aware of it. The sights and sounds we have seen and heard are indescribable. Parts of this Normandy are still untouched . . . but where the fighting has been bitter, the devastation is unspeakable and appalling. I've driven for miles through battle-scarred country . . . feeling just sick inside at the stupidity of it all. . . .

With love, Mandie

Beatrice Cockram of Seattle, Washington, served as a hospital recreational worker for the American Red Cross from December 1943 until the end of the war. She arrived in Cowglen, Scotland, in January 1944 to serve with the Fiftieth General Hospital. Less than six weeks after the Normandy invasion, she went with the Fiftieth General Hospital to France. Her letters to her sweetheart and future husband, Frank Jordan, who worked as an electronics engineer at Boeing Aircraft in Seattle, provide an eyewitness account of the monumental events that took place in France during the summer and autumn of 1944.

[Somewhere in France]<br>July 17, 1944

Dear Frank,

You couldn't be more surprised to find me here than I am to be here. It all happened so quickly. The whole episode is so unreal—so is this war. . . .

Following the packing is all the other regimentation of transportation. More than once I have laughed heartily at myself because I joined the Red Cross to avoid regimentation—only to find myself up to the armpits in it. . . .

The trip across the Channel was like a dream with a big barrage balloon tied to the mast of our ship. . . . Our landing was tremendously thrilling with blue, blue waters and the white wakes of the boats. I couldn't realize we were going to war, but it began to sink in.

Best love, Beatrice

[Somewhere in France]
Dear Frank,                                        July 23, 1944

I am aware of a definite need to get this whole business of War over with and start this great army of Americans going home again. . . . Our nomad life is really shaping up. I feel very much at home in my O.D. [olive drab] fatigues, leggings, etc., carrying my old reliable pig-iron basket (alias helmet, bathtub, laundry tub, Sunday hat, etc.) on my arm. . . .

Best love, Beatrice

[Somewhere in France]
Dear Frank,                                        July 31, 1944

One day I went out on the "mail run" to see one of our teams on "detached service." Some of the doctors and nurses have been working desperately hard. The in-take of patients isn't constant but comes in bunches. Conditions are rugged in field hospitals. To walk down a ward virtually made my heart bleed to see how much was to be done just washing faces and giving drinks of water. Their nurses are too busy with the vital blood plasma, penicillin and sulpha treatments. . . . I had the chance to talk to several pre-operative patients, while they were waiting. They were young enough to be in good spirits— damned glad to be still alive, it seems.

Best love, Beatrice

[Somewhere in France]
Dear Frank,                                        October 16, 1944

. . . Our recreation tent, office and craft shop are away to the east in the midst of the traffic. We really have a pleasant little village across the road from the administrative tents. Village includes the post office, post exchange, barber shop, Coca-Cola tent, and our activities. . . . The big enlisted man's tent has as much white paint as I have managed to find because it lightens it up so much. Except for the top surface, the ping-pong table is white, writing desks and stools are white as are bookcases trimmed with blue felt. There are lots of splashes of red— low red shelves, red stools and red pillows on white stools. . . .

Tossed against this background are a number of pertinent items of the American way of life. . . . In the craft shop, we have long table and benches and stools all rust color. My little niche of an office is in green and rust with yellow wooden shoes for bookends. . . .

Best love, Beatrice


Beatrice Cockram returned to the United States in September 1945 and married Frank Jordan. They had two children. She earned a master's degree in library science from the University of Rhode Island and worked as a librarian for a number of years. She now lives in Kirkland, Washington.

AILEEN HOGAN, a graduate of the Columbia University School of Nursing, was forty-two years old when she volunteered for the Army Nurse Corps early in 1942. She served with the Second General Hospital, a unit that was inaugurated by the Presbyterian Hospital of the Columbia-Presbyterian Medical Center in New York City. Hogan arrived in Liverpool, England, on July 19, 1942, and remained overseas until August 1945. She was with the Second General Hospital Unit when it went ashore in Normandy on July 25, 1944. Her work with the Second General included assignments with evacuation hospitals at the battle front. She wrote long letters to her family and to her sister, Kitty, in which she included important information about the life of a combat nurse on the European front during the summer and autumn of 1944.[16]

                                           [Somewhere in France]
Dear Family:                                      July 26, 1944
. . . We had a grand trip over, were much more in condition regarding carrying packs and baggage than we were two years ago. The sea

---

16. The letters of Aileen Hogan are located in the Aileen Hogan Papers, Special Collections, Milbank Memorial Library, Teachers College, Columbia University, New York, New York. An unpublished manuscript, "Letters from the Second General, 1942–1945," edited by Ruth M. Lee, is based on the wartime letters of Aileen Hogan. A copy of this manuscript is located at the Milbank Memorial Library.

was like glass. The Navy got us ashore without even getting our feet wet. The traffic on the road is terrific. The poor boys have no privacy. They have settled in along the beach and roads with not a thought of girls even being around, and they are slightly startled when several truckloads of us appear. . . .

The artillery fire in the distance sounds as though a thunder storm were blowing up. . . . Taking a bath in a helmet is not at all bad. Then we wash our socks and undies in the same sudsy water, one rinse in clear water. Life is very simple!

Love, Aileen

[Somewhere in France]

Dear Kitty:                                                    August 5, 1944

Yesterday we found that we were about to "fold our tents like the Arabs" and dash madly after the Army. . . . We are [now] sitting on our boxes waiting for the trucks to get around to us and as there will probably be no free movement once we catch up with the Army, I'm dashing this off now and will mail it from our next stop. . . . I'm so glad we are having this chance to see an Evac Hospital moved and set up, and the pressure of work is something you'd have to see to believe. Plasma, blood, penicillin and sulfa are doing a marvelous job. It is good to be really working. . . .

Love, Aileen

[Somewhere in France]

Dear Kitty:                                                   August 11, 1944

. . . It has been a busy week. Started on a medical row, was sent in an emergency to Sterile Supply, a high pressure madhouse, did 12 consecutive hours of "gloves" at top speed just trying to keep up with demand. . . . Next night was suddenly shifted to gas—gangrene tent, an isolated spot at the end of a field. The first night was hectic, constant admissions, no lights except our small G.I. flashlights. Patients getting plasma and blood. . . .

The penicillin team is an interesting detail. At seven, all the penicillin needed for the first round is mixed and two technicians and one nurse make the rounds of the hospital giving penicillin to the patients. One loads the syringes and changes needles, the other two give the hypos. At the rate of 60 to a tent, one gets groggy. It is an art to find your way around at night, not a glimmer of light anywhere, no flashlights of course, the tents just a vague silhouette against the darkness, ropes and tent pins a constant menace, syringes and precious medications, balanced precariously on one arm. . . .

I hope you can make out this scribble. I had no idea I'd have a minute to write but came back on medical row after midnight and everyone is sleeping.

Love, Aileen

France

Dear Kitty,　　　　　　　　　　　　　　　　　August 16, 1944

. . . Yesterday we were sitting in our tents, swearing at the heat that would not let us sleep when we saw a jeep with one of our boys in it, drive in the gate and four vehicles following him. They had our orders to return to 2nd General. Such a scramble of packing and then off, literally, in a cloud of dust. It is not pleasant riding in France—nothing but devastation—forests leafless, limbless, just straight stark trees— roads all under construction—pontoon bridges over rivers. And along the roads French people—very old women with maybe a small child or two and a few clothes in a wheelbarrow, making their slow way back to their homes. . . .

Here at 2nd General there is a bee hive busyness—all the teams coming in, all full of the things they have been doing—everyone talking—no one listening. Outside it sounds like a construction gang— everything but the welding—Everyone has a hammer and a saw. Cement floors are being laid—ward tents are being set—colored troops are getting our areas ready. German prisoners working everywhere. . . .

However, our mess is good. We'll never complain after the last place we were. How those people do the amount of work they do on the food they get is quite beyond me. . . .

Love to you both, Aileen

[Somewhere in France]

Dear Honeys:　　　　　　　　　　　　　　　　August 22, 1944

. . . I was amused at your referring to my "cheerful" letters. Heavens! Why shouldn't I be—I have two arms and two legs—two good eyes— a family to go home to—and a home in a country that is not a desolate waste, devastated beyond anything one can imagine. Our blockbusters are something I hope you never hear. I do not know how these people stand up against it. And in the midst of it all you see two and three year old children.

I have a young woman in the ward. The village was warned to take to the fields but she couldn't find one of the children and by the time she and her husband and five children got started so had our bombers.

She saw her husband with the baby fall flat on the road, what has happened to the others she does not know. She was the only one the advancing troops found. This is only a small incident compared to the agony of these people. We are as carefree as birds. We are now way back behind the lines in our own hospital.

Working with the team at the forward Evac hospitals was a very exciting and satisfying but heartrending job. They do a marvelous job. No time for any actual nursing, all emergency first aid work—plasma, transfusions, penicillin. We hope that maybe we will get these boys back here and be able to give them all the extras they deserve. . . .

Have to run before blackout. Love, Aileen

The next letter includes information on the recuperation of several nurses who were seriously injured in a jeep accident.

                                        [Somewhere in France]
Dear Kitty:                              September 11, 1944

. . . Evie Elwood is taking her first walk—using two canes and Major Wilcox supporting. She is doing splendidly—says her feet feel like lead. Gerry Kiefer is up in a wheelchair. Last night, they created quite a commotion by appearing in their wheelchairs over at our mess for dinner. The Chef and the boys from the kitchen came over and took them over. The chairs would have to be lifted over the hedgerows. Mearnsey is back to work. Betsy Baker is back in bed with a slight infection in her leg. Nothing serious, but the M.D.'s are taking no chances. . . .

The cold here is terrific. How one ever recovers from a sinus I do not know except that we are all pretty rugged. The rain comes down in buckets, runs in rivers under one's bed. The ward men swish it out with brooms. About every two hours the sun comes out bright and clear for an hour or so. The nights are unbelievably cold, woolens, socks, flannel pajamas, sweater, hot water bottle and four blankets, and still the cold pierces thru. It must be dreadful for the boys in the hedgerows. . . .

Love to you both, Aileen

The staff of the Second General Hospital cared for wounded German prisoners of war as well as wounded Allied soldiers. The next letter provides information about the German prisoners of war at the Second General Hospital.

[Somewhere in France]

Dear Honeys:                                    September 24, 1944

. . . Two more trainloads of patients arrived in the night. All P.W.s [Prisoners of War] and pretty sick. We have five P.W.s working with us in sterile supply. None of them speak English, so my German vocabulary is growing larger. Unfortunately there is not an equal growth on the grammar side. So, I continue to murder the German language.

Hans is the baby. Sept. 3 was his 18th birthday. He is Yugoslav, has been fighting three years. He found a shaving brush and had cleaned it up and brought it in to be sterilized. The other four P.W.s razzed the life out of him, as he has a pink and white baby skin and not a sign of a hair on it.

Yesterday was blood donation day and they were all going to the lab. in the afternoon to give blood. There had been a much heated discussion going on all morning. It was too fast, I couldn't get it, but when the first one came back from the lab, looking a little sheepish, I found out. Seems they had given blood in Germany and had been told that 500 cc was the outer limit of what should be taken. Someone here, probably an American, told them the Americans always gave a 1000 cc, so they were not going to say a word, if the Americans could give a 1000 cc, they could too! But the first one back reported 500 only taken! . . .

Love, Aileen

France

Dear Honeys:                                       October 9, 1944

. . . I am having a bad dream. You walk out of your tent and the grounds are being policed by men dressed as you are, in green fatigues, except that P.W. is printed on various prominent parts. You cut across the field and meet an English soldier in battle dress, beret slanted over one eye. You say, "Good Morning," and he says, "Bon Jour" and you realize that he is Free French in English uniform. You pick up a bundle and start for your ward, an English nurse with a perfect Oxford accent says, "Sister, that is too heavy for you, allow me," and you turn and you find the grey blue Nazi uniform behind you. A P.W. of course, but after a while you get mixed up yourself. . . .

Love to you both, Aileen

Aileen Hogan remained with the Second General Hospital until the end of the war, returning to the United States in August 1945.

After the war, she earned a master's degree at Teachers College, Columbia University. She played a prominent role in the establishment of the American College of Nurse-Midwifery, serving as its first executive secretary. Aileen Hogan died in 1981.

AS THE LETTERS of Army nurses demonstrate, their work in Europe was often quite dangerous. On October 21, 1944, Second Lieutenant Frances Y. Slanger of Roxbury, Massachusetts, an Army nurse stationed with the Forty-fifth Field Hospital near Henri Chapelle in Belgium, was killed in the line of duty. She was one of 201 Army nurses who died during World War II, sixteen of whom were killed as a result of enemy action. The day before Lieutenant Slanger was killed, she wrote a letter to *Stars and Stripes,* which was published in that newspaper in November 1944.[17]

> Somewhere in Belgium
> October 20, 1944

It is 0200 and I have been lying awake . . . listening to the steady, even breathing of the other three nurses in the tent. . . . The rain is beating down . . . with a torrential force. The wind is on a mad rampage. . . .

The fire is burning low and just a few live coals are on the bottom. With the slow feeding of wood, and finally coal, a roaring fire is started. I couldn't help thinking how similar to a human being a fire is; if it is allowed to run down . . . if there is a spark of life left . . . it can be nursed back. So can a human being. It is slow, it is gradual, it is done all the time in these field hospitals and other hospitals. . . .

I'm writing this by flashlight. . . . The G.I.'s say we rough it. . . . We wade ankle deep in mud. You have to lie in it. We are restricted to our immediate area, a cow pasture or hayfield, but then, who is not restricted? We have a stove and coal. We even have a laundry line in the tent. . . . It all adds up to a feeling of unrealness.

Sure, we rough it, but in comparison to the way you men are taking it, we can't complain, nor do we feel that bouquets are due us.

17. Excerpts from the letter of Lt. Frances Y. Slanger appear in Pauline E. Maxwell, "History of the Army Nurse Corps, 1775–1948," Volume XI, "The European Theater Through 1944" (Washington, D.C.: U.S. Army Center of Military History, 1976), pp. 117–19.

But you, the men behind the guns, the men driving our tanks, flying our planes, sailing our ships, building bridges and to the men who pave the way and to the men who are left behind—it is to you we doff our helmets. . . .

Yes, this time we are handing out the bouquets . . . after taking care of some of your buddies . . . when they are brought in bloody, dirty with the earth, mud and grime, and most of them so tired. . . . Seeing them gradually brought back to life, to consciousness and to see their lips separate into a grin. . . . Usually they kid, hurt as they are. . . . "How 'ya, babe," or "Holy Mackerel, an American woman!" or most indiscreetly, "How about a kiss?"

These soldiers stay with us . . . from ten days to possibly two weeks. We have learned . . . about our American soldier and the stuff he is made of. The wounded don't cry. Their buddies come first. The patience and determination they show, the courage and fortitude they have is sometimes awesome to behold. It is we who are proud to be here. Rough it? No, it is a privilege . . . to receive you, and a great distinction to see you open your eyes and with that swell American grin, say, "Hi-ya, babe."

[Frances Y. Slanger, ANC]

Romance was also a part of the battle front experience. On December 30, 1944, at the height of the fighting of the Battle of the Bulge, WAC Sergeant Evelyn M. Zimmerman of Portland, Oregon, married Sergeant Steve F. Mitchick at Lieutenant General Omar N. Bradley's Twelfth Army Group Headquarters in France. The next letter, written by Sergeant Evelyn Zimmerman Mitchick to WAC Captain Rose L. Wagnes in Brookline, Massachusetts, describes the wedding.[18]

[Somewhere in France]

Dear Waggie,                                                    January 10, 1945

Thanks so much for the package of wonderful gifts. It came the day before my wedding, so I wore the stockings for the wedding; they sure are lovely. Can use all the other fine things, too. The ribbons are swell.

. . . Guess I told you we had to be married twice, once in town by the Mayor at 11, then at 4 in the church by the chaplain. My C.O.

---

18. The letter of Evelyn Zimmerman Mitchick is located at the Women's Army Corps Museum Archives, Fort McClellan, Alabama.

gave me away, and my gal friend Mac was maid of honor. The officers gave us a wonderful reception at Junior Officer's day room after the wedding. Rode to and from church to the hotel in a Colonel's car. Even one of the generals came to the wedding. . . .

Oh, yes, the night before the wedding, Mitch's platoon gave us a party, what a time. I shouldn't tell you this, but I got drunk, the first and last time though. The stuff they gave me to drink was awful, champagne, cognac, Black and White, and some other nasty fire water. We were both pretty weak the next day, but managed to survive and haven't touched a drop since. . . .

Love, Zimmie

Army nurse Lieutenant Dorothy Brown, a Euchee Indian from Sepulpa, Oklahoma, wrote the following letter to her family shortly after arriving in France. She was one of several hundred Native American women to serve in the military during World War II. Her letter was published in the April 1, 1945, issue of *The Blue Cross,* a newspaper established for service personnel from the Sepulpa area.[19]

[Somewhere in France
December 1944]

Dearest Dad and Brother,

Here I am on the continent of Europe and residing for the moment in a chateau. I really am living in a chateau but what a barn. It's really a tremendous place and sort of beat up I might add. The Germans had it before and there are some German graves up on a little hill by the chateau. . . . The snow is about a foot deep and quite cold but not too bad. You should see me trying to talk to the French. Sign language only but I do ok. This place is only a stop over till I get to my unit which should be in a few days. . . .

[Love, Dorothy]

Much-needed back-up support for the invasion forces was provided by the many uniformed women from the United States who

---

19. The information on Dorothy Brown and a copy of *The Blue Cross* are located in the Samuel W. Brown Papers, Archives and Manuscript Division, Oklahoma Historical Society, Oklahoma City, Oklahoma. Because Native Americans did not serve in segregated units, it is difficult to determine the exact numbers of those who served in the military. See Allison R. Bernstein, *American Indians and World War II* (Norman, Oklahoma: University of Oklahoma Press, 1991), 40.

served in England during 1944 and 1945. To take one example, WAC Private Maud Turner Cofer of Atlanta, Georgia, served with the Motor Transportation Corps in England, where she drove trucks and worked as an automobile mechanic. In a letter of November 5, 1944, to her parents, she told them about the "dirty work" and "grime" of her job. In the same letter, she noted that "the garage has responded wonderfully to what work I am able to get done [but] lacking brute strength . . . I do not keep the place as well as it should be kept."[20]

WACs in England performed important duties as postal workers, helping ensure that treasured letters from home reached the troops as quickly as possible. During the spring of 1945, the more than 800 black WACs of the 6888th Central Postal Battalion, the only black WAC unit to serve overseas during World War II, were sent to England and assigned the responsibility of redirecting the mail to the estimated seven million United States personnel stationed in Europe. The women of the "Six Triple Eight" worked around the clock, seven days a week, at their duty station in Birmingham, where they broke all records in the speed with which they redirected morale-boosting mail to the troops.[21]

Army nurses and Red Cross hospital workers in England also provided crucial backup support for the fighting forces in Europe. During the summer and early autumn of 1944, Army nurse Marjorie LaPalme of Greenfield, Massachusetts, worked at the 168th Station Hospital in England, where she helped care for wounded soldiers who had been evacuated from the battlefields. In a letter to her parents dated July 18, 1944, she expressed her frustration about working in England "when everything is happening across the channel." Nonetheless, the work that she and thousands of other nurses in England performed was crucial to the success of the Allied invasion.

In October 1944, Marjorie LaPalme was transferred to the Forty-first Evacuation Hospital, close to the front lines in Holland and Belgium. Her letters to her parents provide further evidence of the difficult conditions experienced by combat nurses in Europe.

---

20. The letters of Maud Turner Cofer are located in the Maud Turner Cofer Papers, Georgia Department of Archives and History, Atlanta, Georgia.

21. Charity Adams Early, *One Woman's Army: A Black Officer Remembers the WAC* (College Station: Texas A & M University Press, 1989), 151. Martha S. Putney, *When the Nation Was in Need: Blacks in the Women's Army Corps During World War II* (Metuchen, N.J.: The Scarecrow Press, 1992), 99.

On the move in Holland

Dearest Mom, Dad and Kids,                    October 22, 1944

I shall make this into sort of a diary, Mom. I am on the move. . . .
We are attached to the Ninth Army. We crossed in a small ship—the
stormy channel and had to climb down over the side on rope ladders.
That little tiny landing boat looked so tiny way down there and I am
frightened of height, but we had to go down. We were landed on
Omaha Beach which shows evidence of battle. A jeep took us up to
the top of the cliff. We saw the bombed out German bunkers. Every
inch was covered with our boys' blood—that hard fought battle area.
The area was a sea of deep thick brown mud. The little jeeps almost
seemed to disappear in it. We were put in an empty tent and there we
stayed for the rest of the day. . . . That night they piled us in a truck
and we were on our way. . . . We went through St. Lo twice. Many
things were bombed to the ground except the church steeple. . . .
What a dreadful ride that was—cold wind blowing into the G.I. truck.
It rained and the tarpaulin leaked. I could hardly bear to sit. Those
wooden benches were appalling. We stopped at Le Mons for coffee
and doughnuts. The Red Cross had established a canteen and club
there. We tried to clean up a little, washed our face and hands anyway.
. . . Now we are living in a chateau on a hillside recently vacated by
the Germans. We have no water, fires, or beds. Mess hall is a mile
and a half away. We feel dirty and crummy. We have to sleep in
our clothes and have no water. Ugh! I am frozen stiff sitting here
writing this. . . .

November 6, 1944

Left the chateau in the back of a G.I. truck. . . . We pass the time by
singing, everywhere we ride we sing, on trains, trucks, hay wagons. It
passes the time. Everyone of our Allies must be sick to death of "I
Am Working On The Railroad," cowboy songs and anything else we
could think of to sing. Many would join in with us. In Paris, we attracted crowds. The French would surround our truck and gape at us.
I guess we did look pretty rugged, combat uniforms, helmets, dirty
faces. . . . That afternoon we left on a hospital train going to Liege,
Belgium . . . and once again on the back of a G.I. truck into Holland.

To-night we are in an (105th) Evac. Hospt. still waiting our assignments. We are sleeping on the floor on litters but we have had heavenly shampoos and showers, and sleeping in p.j.s instead of our
clothes. . . . Holland is a clean little country, what is left of it, of
course. I guess I didn't tell you these buzz bombs are coming overhead
very frequently. They make a buzzing sound and when you hear a
whistle, look out below. It was terrible in London before we left.

Many people were killed including Americans. The British certainly had a lot to put up with these past four years. When we were traveling here via truck these bombs just seemed to follow us, so frightening. . . . It's extremely noisy around here, lots of anti aircraft guns and shells popping off.

Guess what, the other morning we were interviewed by the press. They sent jeeps for us to take us to their hotel. Two women correspondents asked us questions and we had a lovely time. One of the women is from Boston. She said the interview would be in the Boston *Herald* sometime.

November 15, 1944

Finally I can unpack. We were at the 105th Evac. for ten days. Now I am at the 41st Evac. This is another monastery . . . an ideal set up for a hospital outside of Maastricht, Holland. Tonight I am on night duty, for the next two weeks. We work 12 hour shifts. . . . This hospital is under a roof for the winter, but moves frequently following the Ninth Army. Many buzz bombs go over head—sound like a motor boat. We hold our breath until they pass. I hate it when we are in bed. . . .

I have just put the lights out in my ward. The poor kids sleep on cots with army blankets. Terrible wounds they have. I have a lantern on my desk. Thank goodness the boys are all sleeping with their shots and pain medication. They can forget their pain and anguish.

This huge monastery was used by Nazis for the Hitler Youth out in the country here. . . . On Thanksgiving Day I, being on night duty, slept all day, but we had a lovely big turkey dinner and the mess hall was decorated with flags and banners. . . . wore a skirt and stockings, hadn't worn those for months. . . . We can take showers twice a week—rest of the time we carry hot water up and take sponge baths. . . .

Love to all, Your loving daughter, Marj

Writing from her vantage point with the Forty-first Evacuation Hospital in Belgium, Marjorie LaPalme described the prelude to the Battle of the Bulge.

[Somewhere in Belgium]

Dearest Mom, Dad and Kids,                December 7, 1944

. . . Three years ago today, Pearl Harbor Day, I was working on 2cd floor north, Maternity, when I went into a patient's room and heard the news of P. Harbor. Three years I have been over here. . . .

The buzz bombs have stopped for awhile but enemy planes come over and then the ack ack guns start. We see flashes and explosions all along the horizon. We are very close to the fighting.

Everywhere through all these countries we see the havoc of war. Piles of bricks and rubble, streets and streets of it, once homes and public buildings, bridges bombed. Our engineers replace them with pontoon bridges. Not just a few homes, but the entire city is gone. Unbelievable. . . . It seems it will take them ages to build up these cities and towns again. Where are all the people? We cannot fraternize with the enemy but we don't see them either, except prisoners. . . .

We went into Volkenburg with a group last night. I was so scared there were so many enemy planes over head, strafing, anti-aircraft guns, mortar shells all around us. The sky was lit up. Our M.P.s stopped us dozens of times to look at our dog tags and identification, looking for German paratroopers who were infiltrating behind our lines. I was so glad to get back. What a scary night. . . .

Later, December 8

. . . It was a terribly exciting night last night. So much activity all over the place. Many enemy planes overhead. We could here them strafing, explosive and mortar fire continuously all the night long. Reports are that the Germans have broken though our lines. No one can go out but what they are stopped by our M.P.s, and asked questions about American ball players, states, presidents, etc. They have caught many Germans in American uniforms. It is frightening. The weather is frigid cold and the snow is deep. My heart goes out to our poor boys out there in the dark and cold.

We were ordered to pack every thing up and sit tight. It being a good possibility that we would have to move quickly as a result of the German Ardennes offensive. All excess equipment was taken back and stored in tents and personal stuff was all ready and stored. It was a tense and worrisome few days I can tell you. Thank God we weren't forced to move and we continued our work.

[Love, Marj]

[Somewhere in Holland]
Dearest Mom, Dad and Kids,                    January 10, 1945

. . . Another nurse and I went to a Red Cross Club dance with a couple of Majors. . . . We had plenty of coffee and doughnuts. It closed at 10 p.m. so we went to another 9th Army Club, top floor of a department store. That place closed at eleven so we had a very enjoyable evening all in all.

Thank God—our boys have pushed the Germans back to their original lines which they started. They certainly put up a great stand— great fighters such heroism. We have seen these kids, tired, dirty, sad, seeing their buddies mown down, but they always have a smile for the U.S. Army nurses, always willing to help us—wanting to talk to us about home and families. Grand kids. I love them all! like my own brothers! When I came over here I was just 21—among the youngest— Now these kids are coming in at eighteen. . . .

All my love to all. Love, Love, Love, Marj

[Somewhere in Germany]
Dearest Mom, Dad and Kids,                        April 28, 1945

. . . I am still on nights. We are busy enough. Can you imagine 600 operations in eight days' time. These doctors are fantastic—I have nothing but praise for this hospital. The doctors and the enlisted men are so dedicated. . . .

We had a nice memorial service for Pres. Roosevelt, April 26th. What a terrible thing to happen to him. I went over to supper at 12:30 a.m. and we heard it on the radio. We thought it was German propaganda, it was so hard to believe. We were speechless for a few minutes. Too bad he couldn't have lived until the end. Yes, the Russians got to Berlin before us. . . . I am sure the Germans will be sorry the Russians reached Berlin first—they will not spare them at all!! . . .

Love to all, Your loving daughter, Marj

The announcement of V-E Day, May 8, 1945, marking the end of the war in Europe, brought forth much rejoicing. Yet victory celebrations were muted by the knowledge that the war in the Pacific had not yet been won.

[Somewhere in Germany]
Dear Mom, Dad, and Kids,                        May 13, 1945

All over but the "Going Home!" Hurrah. How did you celebrate back home? It was very quiet up in the field where we were. We had venison for meat when one of the boys shot a deer in the woods. That was the extent of our celebration of V.E. day. . . . The cease fire order had been given in our sector 2 days earlier on May 7 so we thought that was it but it was May 9 when the unconditional surrender was signed at Rheims. We did have a V.E. party but there was nothing but a dirt floor to dance on. However, we had a good time anyway. The music was good!

It does seem funny not to hear the roaring of guns—the ack acks—the buzz bombs and great bombers going over hour after hour at night.

No more helmets either—overseas caps and lights on—No more black outs! It all takes getting used to after so many years of war. . . .

We haven't heard what our future is yet. How is everything at home— Poor Roy still out in the Pacific where bitter fighting is still going on. . . .

All my love to all, Your loving daughter, M.

Marjorie LaPalme returned to the United States in the autumn of 1945 and worked as a maternity nurse in Greenfield, Massachusetts, for the next five years. In 1950, she married Wilfred Faneuf. They had five children. After the children reached school age, she resumed her nursing career. Marjorie LaPalme Faneuf lives in Greenfield and is an active member of the local senior citizens' center where she tap dances with the "Golden Dancers."

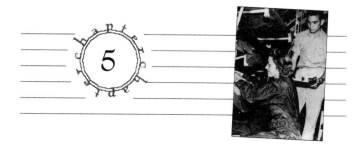

# The War Against the Axis: Japan

THE WAR AGAINST JAPAN encompassed a huge geographic area in which horrendous fighting took place. Attacks on coral atolls, combat in dense jungle, the immense bombing offensives against the Japanese homeland, the movement of supplies to Allies in China, and the recapture of areas that, before the war, had been under the control of the United States demanded great mountains of supplies as well as huge cadres of personnel.

When the Japanese attacked Pearl Harbor and the Philippines on December 7, 1941, Army and Navy nurses were present to care for the casualties of war. The 119 military nurses stationed in Hawaii worked around the clock as they tended to the wounded. Navy nurses aboard the hospital ship the U.S.S. *Solace,* anchored at Pearl Harbor at the time of the attack, attended to injured military personnel who were rescued from the waters of the lagoon. Red Cross workers as well as civilian volunteers also nursed the wounded.

After the fall of Guam on December 10, 1941, five Navy nurses were captured by the Japanese. Following six months of internment in Japan, they were repatriated. When Corregidor fell on May 6, 1942, sixty-seven Army nurses serving in the Philippines were captured and imprisoned at the Santo Tomas Internment Camp, where they remained until the end of the war. In addition, eleven Navy nurses stationed in the Philippines were captured and incarcerated at the Los Banos Internment Camp until February 1945.[1]

1. *White Task Force: History of the Nurse Corps, United States Navy* (Washington, D.C.: Bureau of Medicine and Surgery, U.S. Navy, 1946). Robert V. Piemonte

Throughout the war, Navy nurses were stationed on hospital ships sailing the waters of the Pacific, where they cared for the sick and the wounded. In addition, Army nurses stationed at hospitals in Australia, New Zealand, the Fiji Islands, New Caledonia, the New Hebrides, and New Guinea received war casualties evacuated from the battle fronts. Because of the fierceness of warfare in the Pacific, however, military commanders prohibited nurses from combat areas until they were "secured."

The first of 5,500 WACs arrived in the Southwest Pacific in May 1944, where they performed administrative and office work in wooden-floored tents in outposts such as New Guinea and, later, the Philippines. In July 1944, WACs arrived in Calcutta to participate in the China-Burma-India (CBI) theater of war. By mid-1944, as many as 1,000 Red Cross women were also serving at Pacific and CBI outposts.[2]

Originally, WAVES, SPARs, and Women Marines were not permitted to serve outside the continental United States. In 1944, however, Representative Margaret Chase Smith introduced legislation in Congress allowing Navy women to serve in Alaska, Hawaii, and the Caribbean. When a contentious congressman suggested that these women would find hardships that no American woman should have to endure, Smith tersely replied, "In that case, we'd better bring all the nurses home." The bill was approved, and early in 1945, a few WAVES, SPARs, and Women Marines were assigned to duty in Hawaii and Alaska.[3]

The letters in this chapter were written by uniformed women, stationed outside the continental United States, who helped bring about the defeat of Japan. Reading these letters enriches one's understanding of the many important contributions United States women made to the winning of World War II.

and Cindy Gurney, eds., *Highlights in the History of the Army Nurse Corps* (Washington, D.C.: U.S. Army Center of Military History, 1987), 14.

2. Mattie E. Treadwell, *United States Army in World War II, Special Studies, The Women's Army Corps* (Washington, D.C.: Department of the Army, 1954), 410, 464. Foster Rhea Dulles, *The American Red Cross, A History* (New York: Harper & Brothers, 1950), 471.

3. Frank Graham, Jr., *Margaret Chase Smith: Woman of Courage* (New York: The John Day Company, 1964), 41.

ARMY NURSE BLANCHE M. "RUSTY" KIERNAN was stationed at Hickam Field at the time of the Japanese attack on December 7, 1941, and continued to work in Hawaii until the end of the war. (For other letters by Kiernan, see pages 13–16.) The letters she wrote to her mother in the months after December 1941 describe the routine of hospital and nursing work in Hawaii.

[Hickam Field, Territory of Hawaii]

Mom Darling, February 28, 1942

. . . Our new duty roster for the month of March went on the bulletin board today and I have been put in charge of the operating room. I'm glad that it is only for a month, while Lee [Blanche's roommate] is on night duty because it is a lot of responsibility besides!!!! Besides I'm trying to learn anesthesia too, but the experience will be good for me and I will try and do my best. We now have ten nurses and seven boys in the operating room. What a headache! This is Lee's first night duty since May and she was quite happy about the whole thing.

The ban on liquor was raised here Tuesday and now I think it is being rationed. Anyway, Tuesday Elma introduced Revella and me to some officers she knew who haven't been here very long and who are connected to a tank outfit. They took us to see their camp and tanks and we sat outside their tent and talked. They wanted us to see their swimming pool so we rode a short distance through the woods in a jeep. I got to drive the jeep to the camp and it was lots of fun and we all had a good time. . . . The next thing on the list will be a ride in a tank, and that will be a thrill. . . .

Oodles of love, Blanchie

[Hickam Field, T.H.]

Dearest Momsie, August 2, 1942

. . . I'm anxious to hear more about your prospective position. Will you have to ride the street car and bus every day to Ft. Logan? I hope it won't be too hard on you but it is good of you to want to do your part. It seems there are so many things we can all do if we will, such as being conservative, saving tin, rubber, paper, etc. We have certainly done a lot of improvising in the operating room and hospital. It is good training for all of us. . . .

Love and kisses, Blanchie

[Somewhere in Hawaii]
Aloha Podner,　　　　　　　　　　　　　　December 7, 1942

A year ago today I wasn't sitting down to leisurely write a letter, eh what! The people who buy bonds here in the territory today have them stamped especially with Pearl Harbor on them and the demand was so great and the lines waiting for bonds were so long that the time has been extended to Thursday noon for this special bond. Isn't that wonderful?

Guess what! I'm on duty at a civilian hospital now giving anesthetics to O.B. patients. The Army has loaned me out for two months because they are short of nurses here. I was told about it Thursday and moved Saturday morning. I watched a delivery Sat. night and one Sun. A.M. and gave gas to a mother this A.M. I think I'm going to like it very much altho I haven't had much experience in this sort of thing. . . .

The nurse's home here is very nice and I have a room to myself with a good bed, dresser, bedside table, chair and floor lamp. The nurses are nice and friendly and there are nine other Army nurses here on general duty who commute from one or the other Army hospitals. . . .

[Love, Blanche]


With the collapse of the U.S.-Filipino military effort in the Philippines in the spring of 1942, large numbers of American service personnel were evacuated to Australia, where General Douglas A. MacArthur, former commander-in-chief of the Philippines and Supreme Commander of the Southwest Pacific Theater, established his headquarters. As General MacArthur planned for the Japanese campaign, troops and supplies from the United States were directed to the Southwest Pacific. The following letter was written by an Army nurse stationed in Australia.[4]

[Somewhere in Australia
Late 1943]

Dear Editor of the *American Journal of Nursing:*]

Our first hospital was in tents, with wooden floors, which never looked any cleaner after scrubbing. We began to learn to improvise and economize with whatever was on hand. . . .

On this tour of duty, I really began to know the Australians for I

---

4. Reprinted with permission from the *American Journal of Nursing* 44 (March 1944): 290. Copyright 1944 The American Journal of Nursing Company.

met a family that is an Australian edition of my own. I loved to be with them and it was wonderful to be in a home, with a fireplace, and rugs under my feet. They have saved up their monthly ration of gasoline to take me sight-seeing.

One day, we went out to a sheep station and I rode a horse for the first time in years. We saw all the new little lambs and were fed an enormous tea. We have all learned to talk Australian, which is not at all the same accent as English.

Although I might have chosen to stay at home and help the war effort there, I wouldn't trade my present job with anyone else on earth and I hope they don't send me home until the war is over for good. I like to be in the thick of it, have my information first-hand, and hear another American say, "Gee, it's good to talk to an American girl!"

Second Lieutenant Mary Mixsell, ANC

Martha Alice Wayman of Fairmont, West Virginia, joined the WAAC in 1942. After receiving her commission in 1943, she was stationed at Fort Oglethorpe, Georgia, where she held a variety of duty assignments. On May 12, 1944, she arrived in Brisbane, Australia, aboard the U.S.S. *West Point* with the first WAC detachment assigned to Australia. In late June 1944, she was transferred to New Guinea, where she worked as a mail censor. She was one of the first WACs to be assigned to New Guinea after the eastern portion of the island was retaken from the Japanese during the early months of 1944. In January 1945, Wayman was sent to Leyte Island in the Philippines, less than a month after that island was secured. From October 1945 until her discharge from the WAC in May 1946, she served in Japan. Her letters to her mother provide a good introduction to the life of WACs stationed in the Southwest Pacific.[5] (For other letters by Wayman, see pages 243–245.)

                                              [Somewhere in the Pacific]
Dear Mother,                                              May 5, 1944
. . . It certainly has been some trip, and to think that Uncle Sam is paying for us to go. Of course, we have to work while on board. Either guard duty or censoring. Since our detachment is going out as censors we do the censoring for the entire ship, except the naval personnel.

---

5. The letters of Martha Alice Wayman are located in the Martha Wayman Papers, United States Army Military History Institute Archives, Carlisle Barracks, Pennsylvania.

And, it is quite a job too. But, except for a few things, this isn't really much different from a pleasure cruise. There are a few things such as black outs, no deckchairs, no stewardesses, rather crowded rooms, guards, and other such things that remind me continually that we are not on a pleasure trip. But it really is a nice trip.

It certainly was quite an occasion when we left the states. When we left the camp, we had a band to play some farewell tunes. Then we went to one of the ports on the west coast where we got on the boat. It gave us all a funny feeling as we walked up the gangplank.

We were all out on deck as the ship left the harbor and we stood at the rail and watched the country going farther away. . . . The weather has been grand, but hot. When we left the country and got outside the harbor, the ocean got a little rough and it had an effect on some of the passengers. . . . We had our first meal on board a few hours after we started. Quite a few didn't go, but I was feeling ok so I went down. The meal was swell—as all our meals (2 a day) have been. . . . One thing different from peace time travel on a ship is that we have fresh water turned on only twice a day. Then is the rush for the water. We manage to get fairly clean, even though the water is cold. . . . You should see the ocean! It really is blue—the bluest blue I've ever seen. . . . And the blue is topped often by white foam. Yesterday, the blue was a smooth glassy one. In some places it looked almost thick enough to cut—almost like gelatin. . . . At night the stars are extremely bright. . . .

Until later, Martha Alice


                                                    Somewhere in Australia
Dear Mother,                                              May 13, 1944
By the time you get this you will have received my "canned" cablegram, so you already know where I am. Surprised?

We certainly have done a lot of interesting things recently. We are quite popular (or something) over here since we are the first to arrive. It's really lots of fun. . . . Of course, I can't tell you anything about the trip. . . . You have seen, or will see, pictures of us disembarking. How do you like our outfits? Those helmets certainly are cute, aren't they? And our musette bags on our back! Mine weighed about 15 pounds. I guess the whole trip was swell even tho we worked from 10 to 3 censoring mail. It certainly was good to see land, after days and days of water, water everywhere. . . .

Of course, our coming was a military secret so we didn't have a big crowd. But we had a welcoming committee, newspaper reporters, photographer, etc. . . .

This is all for now, Martha Alice

Australia

Dear Mother,                                                    May 17, 1944

. . . In my letter yesterday I told you what to send me. Add to that
the white blouse I sent to you. I hear we will wear slacks quite a bit
so if they issue them to us I'll have to wear my khaki shirts and insignia. If they don't issue them, I'll have to order some. The nurses have
their own blue and O.D. [olive drab] slacks issued, so we may, too. I
don't need any identification bracelet, I guess, because I wear my dog
tags and a bracelet is in my way. . . .

Has anything been cut out of my letters? After I write something,
I find out that we aren't supposed to say that. Everywhere we go they
change their minds about what we can say. . . . I think this is all for
now, especially since it's time for mess.

Martha Alice


Australia

Dear Mother,                                                    May 18, 1944

. . . As for my not being safe here, forget it. They have rules for the
nurses about not going out with dates unless there are at least two
couples together. They are much more strict over here. Besides there
won't be anywhere to go if we are moved from this town to an army
post. Right now we are near one of the cities and we can go into town.
We go in groups of 3 or 4 and often more. There isn't any danger
here. They sell beer only for about an hour, and it is almost impossible
to get anything any stronger. Everything like that is rationed, even to
Army personnel. . . .

They tell us what we can't say [in our letters] but they always wait
too long, it seems to me. We have lectures on censorship every time
we get to a new place and each place has different regulations. The
ship's rules were the strictest. Our letters are censored by ourselves by
placing our signature on the envelope. Then they are subject to censorship by the base censor.

Our quarters here are very nice and everyone has tried to make us
comfortable. We live in little huts—four to a hut. The latrine and
laundry are a little distance from our house but they are very nice. We
have hot water in the showers and laundry tubs. They even have a
P.A. system that plays recordings for us in the area. . . .

Well, this is all for now, unless I think of something else.

Martha Alice

[Somewhere at sea]

Dear Mother, etc.,                                    June 25, 1944

Now, here I am, back on a ship, headed for—you all guess where. I'll let you know when I get there. This ship is quite different from our other one—much smaller and older. Our quarters are rather crowded, but for the few days we will be on board, it isn't bad. The weather has been pretty rough, and the ship is rolling and tossing quite a bit. It's fun, though, but I'm sorry we didn't fly, as we were supposed to. . . .

I'll write again when I get settled, which should be very soon.

Martha Alice


                                                      New Guinea
Dear Mother,                                          June 27, 1944

Well we are finally at our destination—at least we are here even tho we aren't settled quite. . . . We got in last night and it was late when a couple of us got out here. We were the baggage detail and had to stay with the ship and see that all the bags got off all right. The reason we were chosen was that we were the last two on the alphabetical roster. . . . But, we really had fun waiting around till the baggage was located onto the trucks and then following the trucks with our jeep. The ride to camp was swell—nice and cool, although dusty in spots.

The island is very pretty. As we approached, we could see the hills and mountains all green and pretty. . . . Our camp is rather isolated from everything and very much fenced in with guards and everything. Not guards, just M.P.s to keep everyone out who isn't supposed to be in and to keep everyone in who isn't supposed to be out.

And you should read the regulations regarding going out. We have to go out in pairs, or preferably two couples. And the fellows have to inquire at the gate for a certain WAC and the WAC has to have the name of her escort, his outfit, rank, and all sorts of stuff on a roster at the main gate in order for him to get in. And, of course, there is no parking between here and the final destination, which also has to be given. We get late passes till midnite three nights a week, other nights we can be out to ten o'clock.

And you should see our dress uniform! It really is something. For work we wear trousers and cotton shirts, and when we dress up we wear the same things plus leggings. This is to keep mosquitoes away. Quite fancy.

This outdoor life is really something. We have running water and electricity. The showers, etc., are all out of doors and no hot water but this weather is rather hot, as you can imagine. The water isn't really cold—just cool.

Our barracks have cement floors and the big room is divided up into little rooms with partitions—all very open and nice. There are three of us in our room. We don't have space to put anything except on little ledges around the room. We eat our meals out of our mess gear and wash out our own in boiling kettles of water outside the mess hall. Everything is quite open. . . .

At night we all have to sleep under mosquito nettings to keep not only mosquitos away, but anything else that decides to come in. Under that we feel very safe. Everything is screened in, of course, but sometimes bugs get in through the open door and maybe through cracks in the walls. . . .

Martha Alice

Martha Alice Wayman censored the mail from her post in New Guinea. Other WACs sorted and redirected the mail. Still others were assigned to office and supply jobs. Orders for new materials and the multitude of forms that governed their use and distribution created a large niche for WAC personnel. Indeed, Port Moresby and Hollandia, New Guinea, became important rear echelon bases for the Allies. The next set of letters was written by Frances Harden, a WAC supply sergeant stationed in New Guinea. The letters were sent to her friend, Private Carmen Barnes Skilling, stationed in the United States.[6]

                                               Somewhere in New Guinea

Dearest Carmen:                              September 9, 1944

At last I'm here safe and sound. The trip over was delightful. We had such a good time and were allowed many more privileges than I ever expected. . . . War seemed a long ways off but now that we're here we're face to face with the real problems—dealing mostly with S.O.S. [Services of Supply].

This is a beautiful place—what with the Coconut trees all around, the ocean only a block away, and every thing just as you expected a South Sea Island to look like with, of course, the exception of the military setup which is much better than we dared hope for in this area. The Engineers have done a beautiful job of making this place liveable for us and they deserve a heck of a lot of credit. . . . I like it tho' and to me it's the real Army. I'm glad I'm not in some civilized place to live in a hotel etc. You'd never guess that less than 5 months

6. The letters of Frances Harden are located in the Women's Army Corps Museum Archives, Fort McClellan, Alabama.

ago this very site was nothing but jungle land. Just to see the happy faces of the G.I. Joes as we came ashore was enough to make us glad we came. Some of 'em had never seen a WAC even tho' there have been a few stationed here for some time. Some of these guys have been over here for 2½ years and that's a long time in this section. We hope to see some of them released to go back to the States because of our arrival. That, indeed would make us feel that we'd achieved our aim. . . .

Our hours in the sun are limited, due to the extreme heat until we become acclimated. The days are pretty hot, but the nites are cold. . . . We have to take all precautions against malaria for it seems "little Annie" [the anopheles mosquito] is very prevalent here. I'm convinced that if the Japs or malaria doesn't get us, the coconuts will. They fall all the time. . . .

Love alla time, Frances

Dearest Carmen:

Somewhere in New Guinea
October 12, 1944

Got your precious letter yesterday and was so happy to hear and do so much appreciate all the nice things you said. Friends mean more to me than they ever have, down here—Believe me! . . .

I've been assigned to the Distribution Desk of Stock Records in the Engineering Branch and I like it very much. Been at it for the past 4 weeks. Can't say that I have much more than learned the difference between a screw driver and a diesel motor as yet but at least, I'm trying. These fellows really get a bang out of my ignorance. I come across some big name on these records and I'm convinced that it's Heavy Equipment, but upon being told to look it up in the stock catalogue, I find that it's a 3 inch wrench or some similar item. . . . The fellow whose job I am taking is a peach of a guy. He's so patient and nice to me. Just a heck of a good fellow. . . .

I haven't had on a uniform since I left the states. We have to keep our pants legs and shirt sleeves down all the time as a protection against "Little Annie." Speaking of Little Annie, you should see my atabrine Tan. It's really something, but now I don't look so bad as the brunettes. [Compulsory doses of atabrine were given in the jungle areas as a protection against malaria. It caused the skin to turn various shades of yellow.] They really show up that yellow glow. My freckles have taken on a brilliant golden hue, and I expect even more. . . . If all the atabrine pills we've taken were laid end to end, they'd make a "G String" big enough for Mother Earth. Believe me!

Recently I took a sight seeing tour of the old Battle Grounds here, quite an interesting jaunt and we really trod upon Historical ground—

you can bet on that. We saw where Allied forces made their first beachhead and 'tis said that more than 2/3 of that division was wiped out in that battle. I'm not surprised after seeing the remains. All the palms were mowed down by heavy artillery fire—demolished planes, tanks, landing craft etc. (both enemy and Allied) lying in junk. Craters big enough to bury a house caused by bombs. What a terrific price to pay for a little strip of jungle. This place will really have to pay off big dividends to atone for that loss of life. We saw one of the cemeteries. They're very well kept but sure gives you a weird feeling, but that's one of those things we just don't talk too much about. I only hope those fellows didn't die in vain. . . .

Love alla time, Frances

No matter where United States troops were stationed or how difficult the circumstances, young women and men continued to fall in love. In the next letter, Sergeant Harden describes a G.I. wedding in New Guinea.

[Somewhere in New Guinea]
Dearest Carmen:                                    November 6, 1944

. . . There is so much noise overhead now till you can't hear anything. A formation of planes going over. Just wish you could see them. I never tire of watching them—especially at nite. . . . Sure gives you a feeling of security to see them up there and know they're your own. The fellows here all say that we don't know how to appreciate them until we've been crouched in a foxhole and heard the uneven unsynchronized motors of the Jap Zeros [Japanese fighter planes] along with the whistling of the bombs and then out of nowhere comes the steady roar of our own American fighter planes. . . . No! I don't suppose we appreciate their true value. . . .

We had a big wedding here last Sunday nite. One of our WACs married some G.I. and they really did it up in dark Brown. Chapel, wedding, bridesmaids, groomsmen, music, flowers, candles, etc. I just wish you could have seen it. Everybody was invited. Her C.O. was matron of honor and four of her friends brides maids. They wore evening dresses borrowed from the Red Cross. The Captain (male officer) that she worked for gave her away. The groom and his groomsmen were dressed in khaki uniforms while the bride's gown was fashioned from a parachute given her by a member of an Airborne troop near here. It was made beautifully, and her veil was made from a khaki colored bed net (mosquito). She bleached it white and I mean it was white too. Her flowers were jungle orchids and they were simply gorgeous. The groom's buddies scouted the jungles to find them for her. The altar was also decorated with candles and jungle flowers. One of her friends sang, "I Love You Truly," and "O, Promise Me." Then the male quartette sang, "Love's Old Sweet Song." They used the ring ceremony and afterwards a reception was held just outside the chapel. Her platoon was invited. The post bakery donated a lovely cake. After the reception they were given a jeep and a 3 day pass plus the use of a thatched hut on the beach that was once used as an English mission. I suppose love was in full bloom by that time. They couldn't have picked a more romantic spot. I know that—and everything is lovely now but just wait til his number is called to go back up north. This is when real love stands its test. You'll be seeing the wedding in the news reels soon for a flock of cameramen were present. They'll play it up as a big society roll, I suppose. When you see it you can feel very close to me for I was right there. . . .

Love alla time, Frances


                                                            New Guinea
Dearest Carmen:                                        January 25, 1945

This is my noon hour and I'm staying at the office to try to get caught up on my correspondence. That's the only way I can do it now. We are terribly busy these days. You have an idea what I mean by reading the papers. We have to keep the supplies flowing to those Yanks up there, and they've really got the pressure on the Japs. By the time you get this, they will have marched through Manila and that's the biggest thing that has happened as yet. [Manila was secured on March 3, 1945.] It has been a hard fight but those Yanks can't be whipped. We're all agreed upon that. I've worked till 10:30, 11:30 and 12 every nite for the past week. Maybe you think I'm not whipped down. From 7 a.m. to those hours is a long grind. When I go home at nite tho' I feel satisfied in my mind that I have contributed my bit and I go to sleep without moving till 5 a.m. That's how exhausted I am. Of course, the heat here plays a big part in that—a fellow just came in and said the thermometer by the side of the door to the office here was standing at 128 degrees, and that's plenty warm. Dontcha think?

Irving Berlin and his troop are here to present, "This is the Army," at our theater to-nite. [Berlin actually wrote a special set of lyrics for the popular song, "This Is the Army, Mr. Jones," for the WACs stationed in New Guinea.] Wish you were here to see it with me. They say it's wonderful. They've presented it at several places in New Guinea already. Lanny Ross [a noted singer of the period] is along. It's to be quite a big social event, I understand. Just hope I can get a seat. Irving has been through the office several times and lots of the girls have his autograph, but as you can guess, I'm no autograph hound. My hat is off to those entertainers who give up their lives of ease and comfort to come here and boost the spirits of a bunch of G.I. Joes and Janes—and believe me, I know what entertainment can mean to these fellows who have been over here so long. It's just a glimpse into the past—also into the future and just a taste of home. . . .

Well, honey, I've just been informed that I have a rush job to do so I'll have to quit and get started or I won't get thru' in time to see the big show. I promise a long letter next time if I can possibly find the time. Keep writing, and Be sweet cause I love you.

Love as always, Frances

New Guinea
Dearest Carmen:                                                April 2, 1945

. . . Today, we heard the news that Hitler was dead. [Hitler actually committed suicide on April 30, 1945.] I think it's a darned shame the old goat had the privilege of laying down and dying. Guess he was scared to death when the Russians began knocking on his front door. I still think the Allies were cheated and denied a gross pleasure of doing away with him. On our own Information and Education program this a.m. we were led to believe that the Allies were dubious of that news and wouldn't be a bit surprised at seeing him up and walking around. If that be the truth, then woe be unto him. Guess the German people are a bewildered lot—not knowing which way to turn—but my guess is that some of them will be mighty relieved to get this over—win or lose. The next three days should see the end of German endurance to hold out. They can't go much further—Can they? . . .

Write soon and often, cause I love you.

Your Pal Always, "Fran"

WAC Private Betty "Billie" Oliver of Chicago, Illinois, also served in New Guinea. She worked as a mail clerk at Port Moresby from August 1944 until May 1945, when she was transferred to Leyte Island in the Philippines. Shortly after she arrived at Port Moresby, she met her future husband, Mike Donahue, a sergeant in the Army Air Forces, on a blind date. Despite the pressures of her job as a mail clerk at a Pacific outpost, she wrote enthusiastic letters to her parents about her love for "the sweetest guy in the world."

New Guinea
Dear Family:                                    September 10, 1944

Well, your wandering girl has finally come to rest in her South Sea Island Paradise. I always say that it is not as the movies picture it, but considering everything it is very nice. . . . We have movies every night and eventually we will have dances. So we can gaze at the men and they can stare at us. It has been almost three months since we have actually seen men and several years since they have seen any women. So you know what I mean. . . .

Gods, is it hot here. The perspiration just rolls off you when you just sit in your underwear. Oh, well, I came to do a job and I can stick it out as well as the next guy. . . . This is really an interesting place—coconut palms, grass houses, odd shells on the beach, huge trees with roots that spread for half a block and natives with red hair and rings thru their noses. Really things I never expected to see. Last night we sat on the beach and watched the stars and the waves breaking. Very romantic, only no men. But, I'll work on that after awhile. . . .

Love, Billie

New Guinea
Dear Mom, Dad and Kids:                          October 19, 1944

Don't mean to be neglecting you, but it seems that the days fly by so fast, I don't have time to do a thing. There just isn't enough time in twenty-four hours to accomplish everything. I get up at five, wash, dress, sweep my floor, make my bed, straighten things out, fall out for reveille at 5:30. Do my detail, like cleaning a latrine, or something, eat breakfast, go to work, get off at 10:45, eat, wash my face, listen for my name at mail call, put in a pass for my boy to get in so he can take me. I know you will probably think it couldn't happen in such a short time—but it really has and as far as I'm concerned, it's the most wonderful thing that ever happened to me. Gosh, he's a swell fellow. I know you will like him. In fact, I guess I fell for him because he reminds me so much of Dad. He has black hair and talks and acts just like Dad, tho he is much thinner. When I look at him from the back when he is explaining something (with gestures like you, Dad) it seems just as tho I were home.

We are sure it is the real thing, but there is the possibility that it could be New Guinea so we shall wait til we are both back in the States before getting married. You will see Mike before I get home because he is eligible for either rotation or a thirty day furlough. So he will stop in Chicago for a couple of days to see you and let you put your stamp of approval on him, which I hope you do. . . . Gosh, I'm such a happy kid!! . . . I've seen him every day since I met him. When I work nights he comes out on my half hour break with sandwiches and cold lemonade and stuff. . . . He sure is wonderful. . . . Hope the shock isn't too great for you all. . . .

Your happy Gal, Billie

New Guinea

Dear Family:                               October 23, 1944

. . . When that love bug hits you for sure you really know it. It may surprise you to know that I too have been bit by it. Yes. Ring out the bells—gather a crowd. Your gal is engaged to the sweetest guy in the world. . . . I gotta see him every night that I am off, cause he'll be leaving soon. I will be a lost kid when he goes. I've seen him every day since the third day I was here so I guess you know your gal has it bad. . . .

Your gal, Billie

New Guinea

Dear Family:                             November 27, 1944

I'm sorry about not writing sooner, but it just seems like so many things have been happening recently, I haven't had a moment to myself. . . .

November 28, 1944

. . . I am sitting home tonight for the first time I've been here. My boy left today. I don't know when I felt so bad about anything. But I know this is war and I did have longer with him than I expected so I can't feel too bad. By the way, he's not on his way home, yet—I'm kind of a hoping when he gets there he will be your son-in-law. So keep your fingers crossed for me. I guess you know that that would make me the happiest gal in the world. But, I guess you know that. Sorry, I didn't even tell you his name—it's Michael Donahue (a little bit Irish) he is 27 years old—black hair, blue eyes. Rather skinny from New Guinea at the moment. . . . He is from Vermont and from what I've heard of his family, they are pretty swell people. There were thirteen in the family, but three died. He now has nine brothers and sisters, says we have to at least beat that, but I don't know. If I don't get out of this army pretty soon, I'll be too old to have them. But, I'll sure try. . . .

All the love in the world, Billie

New Guinea
Dear Family:                                      December 1, 1944

. . . Gosh, I sure miss my boy. He has only been gone four days but it seems like weeks. I don't know what to do with myself. . . .

I just can't seem to get myself to do anything but work and come home and wish for my boy. Seems as tho all the girls are rather irritable and cross these days. It is so damned hot we just can't stand it. It is even hot at night now. I've been waking at three every morning and just sitting and waiting for five o'clock to come. We have all been working pretty hard and longer hours because of the rush of Xmas cards so I guess all in all we are just worn out.

Much as I like my job and all the girls here, even tho I know I am, in my small way, doing something to help this mess, I still can't help feel that I WANT TO COME HOME. I know I asked for this and I'm not kicking but gosh! I'm lonesome. Guess I've blown off enough steam to go out with this letter. . . .

There is another big favor I'd like to ask of you. On account of, if everything turns out as planned, I shall be married around Christmas, and all I have are G.I. clothes. Will you please get me a real sexy *nightgown*. You know, the kind a bride would wear? Better get size 40 so it will fit. I can take it in if necessary. I would like a built-up bra effect, blue if possible, silk, crepe or something. No satin. I don't care how much it costs. I just want a real real nice one. Be sure it's long. I'm 5' 8". I sure would appreciate it. Instead of sending that straight mail, send it airmail. You can put it in one of those big brown envelopes and it should get here o.k. I sure hope you can do it. After all, I will have to get married in uniform. I'd like to let my boy know he's got a woman, not a soldier.

Write soon, Love, Billie

New Guinea
Dear Mom and Pop:                                December 19, 1944

Mike's back!!!!! That is the biggest news at the moment. I guess you can imagine how happy I am. He was gone three weeks and I sure was down in the dumps, but now everything is just fine.

We have sent our papers in for approval and they shouldn't take too long so we should be married on or about Dad's birthday. Of course we can't be sure yet when they will get thru, but we sure are hoping for the best. . . .

We have been invited to a native Christmas festival by Harina, our friend. Of course, if we go we can't eat or drink anything they have cause it isn't sanitary, but I would love to go out and see how they celebrate Christmas. It would certainly be something to tell our grand-children. . . .

All my love, Billie

Betty Oliver and Mike Donahue were married on January 4, 1945. In late February, Mike Donahue, who had served in the Pacific for four years, was transferred to the United States. Betty Oliver Donahue remained in the Southwest Pacific until June 1945.

New Guinea<br>April 15, 1945

Dear Family:

. . . Finally heard from Mike, as you probably have too by this time. He arrived home the night before Easter and I guess he's been running around a lot seeing all the people he hasn't seen for four years. He said he is getting awfully tired of visiting people and having them visit him. I suppose life over there is quite different than it was here. He said he has written you. Hope you got his letter. It's pretty hard for him to write, not knowing you, and he never did write much, even to his family. I do hope he has called you tho'.

We had a formal parade and twenty-one gun salute in honor of the President yesterday. It was a great shock to all of us to learn of his death. . . .

I got a letter from Mike's sister, welcoming me to the family. That made me happy, plus the picture of Mike I got. I'm a pretty happy gal again. . . .

Love, Billie

P.I. [Philippine Islands]<br>May 10, 1945

Dear Family:

. . . I went to visit some of the battle casualties at the hospital. Some of those fellows may be crippled all their lives, but they still have a wonderful sense of humor. They are so happy to see us—Laugh at everything we say. We always try to wear something funny to make them laugh. I wore my hair in pigtails with bows on them and you should have heard them howl!! All the girls try to drop in on the wards when they go on sick call— They know how happy it makes the fellows. . . .

We were all issued five pair of women's khaki slacks which has improved morale no end. The kids were really unhappy about their beat up, old clothes. . . .

Lotsa Love, Billie

P.I. [Philippine Islands]
Dear Family:                                    May 20, 1945

. . . I'm applying today for duty in the States—finally got a copy of Mike's orders that I can use. I suppose it will take several months to go thru and then I'll have to await transportation, but I'm not really unhappy here and if I know that eventually I'll get back, it won't be too hard to wait. . . .

Mike is stationed in Louisiana. I guess he must hate it, cause he never wanted to go down south. Don't know where I will be stationed if I do get home on this deal. I'd hate to be in the south too, but if I could be near Mike, it wouldn't make any difference where it was. . . .

Gotta get back to work. Please keep writing and keep your fingers crossed. I still hope to be in the states for Thanksgiving.

Lots a Love, Billie

Betty and Mike Donahue were reunited in Chicago in July 1945. For thirty years, they ran the Donahue family farm in Vermont. They had six children. Mike Donahue died in 1986. Betty Donahue lives in Graniteville, Vermont, where she enjoys visits from her nine grandchildren. She is a founding member of the Vermont WAC Veterans Association.

THE AMERICAN RED CROSS established canteens, recreation huts, and clubs for the many troops who were stationed in New Guinea. Red Cross personnel also served as hospital workers, and there were even a few clubmobiles on the island. ARC worker Jacqueline Haring of Detroit, Michigan, was sent to New Guinea in December 1944 to help establish Red Cross canteens. She was transferred to the Philippines in August 1945. In November 1945, she went with one of the first groups of Red Cross women to Japan. She returned to the United States in January 1946. Jacqueline Haring now lives in Nantucket, Massachusetts, where she is the curator of research materials for the Nantucket Historical Society. The following letters, written to her parents, contain vivid descriptions of her work with the Red Cross in New Guinea. (For other letters by Haring, see pages 238–243.)

[Somewhere in the Pacific
Late December 1944]

Dear Mother and Dad,

. . . I spend all my days on deck talking with the officers and I've heard some of the most marvelous tales of their experiences. Walking out on deck is like stepping out into a strange world where adventure is a chief commodity and tales of heroism are common conversation. Sentiment is present too and we sing songs for men who once sang them and called them favorites. There is poetry and music mixing in the wind and a comradeship that makes it all seem plausible. This is an experience I will never forget and one for which I'll be immeasurably richer.

Christmas was really wonderful altho a little strange. I joined the ship's choir and we sang at three church services. Sunday we caroled all through the ship after taps. Then on Xmas day we sang in the Sick Bay and for three Xmas shows. . . .

Don't worry about me. I've never been more certain that this is right.

J.

[New Guinea]
January 10, 1945

Dear Mother,

I still can't believe I'm so many thousands of miles away from the quiet, green streets of Ann Arbor and the bustle of Detroit. . . . When we disembarked, we were met at the dock by a group of ambulances (the only available form of transportation) and driven about twelve or fifteen miles to our present home. I felt just as if I were walking into a prison camp when we arrived for our compound is just a group of about twenty tents, a shower shack, recreation hall (unfinished) and a mess hall all completely surrounded by a high wire and burlap fence. . . . The command is very strict here about the protection of women and we have a hundred regulations we have to adhere to. No one can go out at night unless her escort is carrying side arms and has authorized transportation. . . . It's really very reassuring to know those M.P.s are there for I guess there are unfortunate things that have happened not here but elsewhere on the island. . . .

One of my roommates is a little older girl. . . . She has been in Greenland for six months and has now come to the other extreme. The other girl is also very congenial and comes from E. Orange, New Jersey. We were able to choose our room-mates here. . . .

Love, Jackie

                                                        [New Guinea]
Dear Mother,                                        February 15, 1945

The canteen is finished! I was even able to take a day off yesterday
and feel that everything would run smoothly. I'm really pretty pleased
with it and everyone has been most complimentary about it. Of
course, it still isn't a thing of beauty but it *is* an improvement and
there have been several innovations. Our bulletin board with the latest
news and all kinds of maps from the Information and Education section seems to be making quite a hit. Then, I managed to get us a
Victrola and four albums of records so we have music going all the
time and that seems to make everyone feel peppier. Our library is
developing slowly but surely and I think we look very gay in all our
red, yellow and blue ammunition paint. . . . I have a bulldozer coming tomorrow to level off the whole area in front of the canteen and
that should make a much better parking area for all the trucks and
every other kind of vehicle driven by my clientele. By now, I've
learned a little more about how to ask for things I need and whom
to ask. It's amazing how willing and anxious people of all kinds and
organizations are to do things to help out. I swear they do much more
for us than we could ever do for them, but they seem to get a huge
kick out of doing it. . . .
Much love, Jackie


                                                        [New Guinea]
Dear Mother and Dad,                                February 23, 1945

I've seen history in the making! —and I can't tell you about it. All I
can tell you is that I've seen the American Prisoners of War who have
been in concentration camps for three years. [Haring is referring to
the repatriation of Americans who had been imprisoned by the Japanese.] I wish I could put into words the inspiration I felt at seeing
them. They were so grateful and appreciative of any little favor we did
for them. They kept thanking us! I would have given them the moon
and stars if they'd asked it, but you could count on their not making
any demands. Of course, they did tell me many stories which I can't
write here but they are things I'll never forget. . . .
    Now it looks as if we may get a permanent building for the canteen
after all. A colonel from the base engineers came to see me Sunday
and said he thought something could probably be done. His construction supervisor came day before yesterday and I gave him the
plans I thought would be ideal and now I must wait to see if they're
possible. Of course I've asked for a lot more than I may get but I'm
hoping a concrete floor, a drainage system, and running water even

though I still have a tent overhead; that would be a 100% improvement. . . .

Much love, Jackie


Faye Anderson of Birmingham, Michigan, worked as a staff assistant for the American Red Cross in New Guinea and the Philippines during 1944 and 1945. After the war, she remained in the ARC, working at Red Cross clubs in Germany with the occupation forces. In the following letter to her family, she writes of the dreadful destruction of Manila and of her pride in the work of the Red Cross in the Philippines.[7] (For other letters by Anderson, see pages 257–258.)


[Manila, Philippines

Dearest Family: Early July 1945]

I am going to try to relate properly the most exciting week of my life so far in the big Red Cross. . . .

Manila, as it is now, is more than I have words to describe. The destruction, starvation, persecution, and ruin is incredible, to say the least. It's a down right crime to have happen to any place what happened here. . . . To put it short, there is literally nothing left of what must have been the most beautiful spot in the world, some buildings dating back to the 1600s. The art, architecture, traditions, etc., that have been lost here will take two generations to regain.

It is all in a state of complete confusion . . . [but] it's a great thrill to be an American and see how our boys have taken hold, working day and night to further this war. . . . The Filipinos look to us like Gods and can't do enough for us as we are the Yanks that saved them from a fate worse than death. As you pass along the streets, even the half starved, dirty, bare little urchins make the "V" for victory sign with their fingers, and yell out, "Victory Joe" as every Yank to them is a Joe. We should be very proud of our nation. . . .

This city is filled with tragic stories and it makes you gasp to hear some of them. The Filipinos really suffered and I'm amazed at the way they are able to take it all. Their city in ruins, half starved, their families killed and tortured in front of their eyes. We all don't know how lucky we are to be Americans. I was talking to a charming Filipino woman who evidently came from a very good family and she said they

---

7. The letters of Faye Louise Anderson are located in the Faye Louise Anderson Papers, Bentley Historical Library, University of Michigan, Ann Arbor, Michigan.

wouldn't have cared if there hadn't been a pillar or post left in the city as long as the Americans arrived.

. . . I will say that the Red Cross is doing one marvelous job here and I'm proud to be in their organization. From ruins they have already got a place started where our boys can get a snack or cold drink, read a book, play some ping pong, meet friends, write letters or just plain relax. Their civilian relief help will go down in history and believe me, all the gals are working at top speed. I'm so glad they put us to work for I love it and we are needed so badly. As I say, every man, woman and child is working like dogs. This is the first bit of real war we've had, but we are entitled to it, for we sweated out a back base long enough. . . .

My Love to You all, Faye

The following letter, written in February 1945 by a WAC stationed "Somewhere in the Philippines," draws significant contrasts between conditions in New Guinea and the Philippine Islands.[8]

                                   "Somewhere" in the Philippines

[Dear Friends:]                                             [February 1945]

This is quite the place! Arrived early in February and, in comparison, New Guinea was the Ritz, no fooling.

Here we live in regular tents, instead of the hut/type, but are fortunate that they have wooden floors. No lights, except candles. Latrines are in tents, community style; also, no lights at night. Showers and washstands are out in the open so that at mid-day one gets a sunburn with the shower—and perhaps a heaven sent shower to boot.

. . . The water situation (as distinct from the RAINFALL situation) is a hit-or-miss proposition. For example, the other night between supper and returning to work, I tried to shower and shampoo. When I was entirely covered with lather, the water stopped and I had to stand about twenty minutes, with my eyes closed, hoping (practically praying) for the water to resume. Finally, it *dripped* through. . . .

The rainfall—well, it's hit-and-miss, too, but mostly "hit." Rains every little while, day and night, whether there is sunshine or not. In New Guinea we'd have a good, hard shower every 24 hours or so, and the rest of the time it was clear. No such thing as overcast days—at least where we were. . . .

---

8. The "Somewhere in the Philippines" letter is located in the Women's Army Corps Museum Archives, Fort McClellan, Alabama.

The place we are at is really a native town—really native, and it's too large to be called a village. Grass huts, no sewage system, no pavements, and mud, mud, everywhere—as well as dust where there is no mud. Most of the people go barefoot. . . . Most of them talk English, some broken English. How the people manage to keep themselves so clean looking is a miracle. They are rather wholesome looking, not emaciated as shown in pictures of Europeans. . . .

It's too bad we don't have a lot of old civilian clothing, or cloth of any kind, they need it so sorely and would rather have an old T-shirt, dress, kimono, sheet, or anything like that than money, because there is no merchandise for them to buy thereabouts. . . .

The trip up here was made in a convoy, which was very interesting, but of course slowed up the travel. We were in a reefer freighter (refrig to you) and slept in half open hatches, in canvas bunks on iron frames, two high—like you see in some of the newsreels of transport— along with endless numbers of roaches for bedmates. Wonderful! The deck latrine (the only one) was something out of this world. A small, oblong shed, dark even in daytime, and pitchy-black at night. It housed a row of 10 or 12 "accommodations," and unless one wore a watch with a luminous dial, there was no way of telling which were occupied; so, upon entering, one sang out "Any vacancies?" and then felt along for pairs of knees, for the first blank spot. As I said, out of this world. The tissue was hung on an overhead rope, and you slid it along to whoever called, a system of systems. . . .

We haven't had any night alerts since I arrived, but just "in case" the camp is dotted with slit trenches and foxholes—a regular obstacle course, particularly when there is no moon. Even in the daytime, one has to beware of tent ropes that trip you up, or tend to decapitate you wherever you walk. All in all, I wouldn't miss it all for a farm, let alone the whole state of Utah. . . .

[Name unknown]


Throughout the bitter fighting in the Pacific, Navy hospital ships, including the *Solace,* the *Relief,* the *Samaritan,* the *Bountiful,* and the *Refuge,* ferried the sick and the wounded from battle stations to hospitals in New Zealand and Australia. Hospital ships also docked in harbors under heavy fire and received the wounded directly from the beaches.

During the fighting at Iwo Jima and Okinawa early in 1945, the Navy began to use hospital-equipped evacuation aircraft to move critically wounded patients to large hospitals at Guam, Pearl Harbor, and the continental United States. Navy flight nurses assigned to the Naval

Air Evacuation Service landed on captured airfields in the Pacific and often began to treat the wounded while they were being evacuated into the aircraft.

The next two letters, written by Navy nurses to the editor of the *American Journal of Nursing*, describe life on hospital ships stationed in the Pacific.[9]

[U.S.S. *Bountiful*
January 1945]

[Dear Editor of the *American Journal of Nursing:*]

Presented in terms of a civilian hospital, the task of the Navy hospital ship would be an impossibility. Imagine receiving approximately five hundred fresh casualties within the space of a few hours! Yet, thanks to planning and organization, it is accomplished smoothly and speedily. Those requiring immediate surgery are treated at once. Other cases are taken to the proper wards, where they are bathed as quickly as possible, and allowed to get some much needed sleep before further treatment is attempted. Few, except combat fatigue cases, require sedation; extreme physical exhaustion seems to overcome pain. Clean, rested and well fed, their wounds promptly and carefully treated, natural vitality comes to the foreground, and convalescence is rapid in most cases. The wonders of plasma, the sulfa drugs, and penicillin could never be better demonstrated than on a hospital ship caring for fresh casualties.

Now, having seen much of the central and south Pacific, life aboard has resolved into the preparations for action, caring for the patients until they may be further transferred, looking forward to mail and liberty, and anticipating homecoming when our tour of duty is completed.

Between trips with patients, we hold "Field Day" [the Navy equivalent of a GI Party]; the wards and special departments are given a thorough cleaning, airing, and repainting if necessary. Supplies are made, packaged, and sterilized and all hands relax, rest, and do odd jobs.

Before we leave the ship, we expect to see further action, and perhaps take part in the final invasion before victory. When back ashore we shall be able to look back at our sea duty and have the satisfaction of work well done.

Lieutenant (junior grade) Georgia Reynolds, Nurse Corps, USN

---

9. Reprinted with permission from the *American Journal of Nursing* 45 (March 1945): 234–35. Copyright 1945 The American Journal of Nursing Company.

The following letter, published in the December 1945 issue of the *American Journal of Nursing,* describes the activities aboard the U.S.S. *Refuge* during the previous Christmas. The writer is probably discussing activity in Leyte, but censorship prevented her from giving exact locations.[10]

[U.S.S. *Relief*
Autumn 1945]

[Dear Editor of the *American Journal of Nursing:*]

Yes, we had Christmas aboard the *Refuge* last year, but it was a far cry from those we had experienced before. . . . We prepared for it just as though we were home with Mom and Pop and the kids. We decorated the wards and wardrooms and planned delicious menus. For weeks the choir practiced Christmas carols and were disappointed when they did not have the opportunity to sing them.

Then came Christmas eve. We arrived at our destination! We had scarcely dropped anchor at 1300 when LSTs [landing ship tanks] were alongside with casualties. We embarked patients with as much speed as possible and prepared to give First Aid to those more seriously injured. All these boys were fresh casualties and many needed intravenous fluids and plasma, morphine, dressings changed, and a good bath or shower. Patients were brought aboard all day Christmas and the day following. For two or three days we worked with all the energy we could muster before things were squared away and the patients showed marked improvement.

Maybe this was not like Christmas at home, but the spirit of kindness, giving, and thanksgiving was there. . . .

Navy Nurse


For United States citizens, the China-Burma-India theater (CBI) was the most mysterious and least known of the principal battle areas of World War II. When the war began for the United States in December 1941, a major conflict had been raging in East Asia since 1937, when Japanese troops crossed the Marco Polo Bridge near Peking (Beijing) and moved into China. By 1939, Japan occupied more than 900,000 square miles of China and controlled most of its major ports and cities.

---

10. Reprinted with permission from the *American Journal of Nursing* 45 (December 1945): 1,063. Copyright 1945 The American Journal of Nursing Company.

After Pearl Harbor, Japanese forces moved rapidly into southern Asia. By late May 1942, the Philippines, Hong Kong, Singapore, the Dutch East Indies, and New Guinea were all in the hands of the enemy. As Japan conquered southwest China and Burma, Allied forces fled across the mountains to India. With the closing of the Burma Road by the Japanese in April 1942, the U.S. Tenth Air Force in India began to fly desperately needed supplies over "The Hump" (the Himalayan mountains) to Allied forces in China. In addition, members of the American Volunteer Group (the "Flying Tigers"), who had first arrived in China in the late summer of 1941 to provide air defense for the Burma Road, were transferred to Kunming, where they engaged in heavy fighting with Japanese aircraft.

Americans who served in the CBI were largely concerned with maintaining a supply line to southwest China, both by airplane over "The Hump" and by land across the Ledo Road. The first American Red Cross workers in the CBI included five women and three men who arrived at Karachi, on the west coast of India, on May 16, 1942. A small number of Army nurses were stationed in the CBI during 1943 and 1944 where they treated American and Chinese troops. The first of several hundred WACs arrived in Calcutta in July 1944.[11]

In the following letter, an Army nurse provides a rather roseate account of her duty in eastern India, where oppressive heat, filth, and a very high malarial rate presented difficult obstacles for nurses to overcome.[12]

[Somewhere in India
January 1944]

[Dear Editor of the *American Journal of Nursing:*]

It is a year since we arrived at our first station, where the hospital was only partially built and there was nothing but jungle to be seen. Both officers and enlisted men worked, and soon, where before had been nothing but bamboo trees and undergrowth, appeared cleared ground,

---

11. Richard Hisgen,"The History of the American National Red Cross, Volume XV, The American Red Cross in China, Burma, and India, 1942–1946" (Washington, D.C.: American National Red Cross, 1950), 5. Judith Bellafaire, *The Army Nurse Corps in World War II* (Washington, D.C.: U.S. Army Center of Military History, 1993). Treadwell, *Women's Army Corps,* 464, 772.

12. Reprinted with permission from the *American Journal of Nursing* 44 (March 1944): 290. Copyright 1944 The American Journal of Nursing Company.

tents, a mess hall, and walks lined with bamboo and filled with stone. Our bamboo bashas appeared as if by magic. Each basha has two rooms and two nurses assigned to each room. The basha is a native hut constructed entirely of bamboo except for the palm leaf roof. Large green bamboo poles are hammered until they become fairly soft and pliable. They are then split lengthwise and laid flat. The natives weave these pieces like mats and use them to form the walls, partitions and floors of the bashas. Rafters, corner poles and joists are made of whole bamboo poles. Other pieces are split and used in strips on the roof to hold the palm leaves in place. (Though they don't always serve the purpose.) . . .

When we first came, we found beautiful purple butterfly orchids hanging from the trees and a profusion of hibiscus blossoms, too. At night we heard the jackals howling, and in the early morning, the monkeys. There were many snakes, but none in our immediate compound. Those who were afraid of cows soon got over it. They are everywhere in India—in the street, in houses, in shops—in fact, when we happened to forget to fasten our door before going out, we occasionally returned to find a cow in our basha, reflectively chewing her cud. . . .

Then, a part of our organization was transferred to a new location. We had a great deal of work and fun in setting up a plan for the care of patients. Many of our patients are not of our own race and this has been a most interesting experience. It is impossible, at times, to teach another standard of living to them. We have had to learn, rather, to adapt ourselves to their customs in some instances, especially in regard to diet. . . .

First Lieutenant Matilda E. Dykstra, A.N.C.


The next letter, written by an Army nurse stationed in western India, depicts the conditions encountered on her side of the subcontinent.[13]


Somewhere in India
[Summer,] 1943

[Dear Editor of the *American Journal of Nursing:*]

Our quarters are very nice and the sand is not blowing as it did before the rains. The girls live two to a room with a shower in each. Ceiling

13. Reprinted with permission from the *American Journal of Nursing* 43 (August 1943): 769. Copyright 1943 The American Journal of Nursing Company.

fans keep the temperature comfortable even when it is very, very hot. Our quarters are spread all over the base with four to twelve nurses living in each.

Our food is very good. We have iced grapefruit juice for one noon-day meal, next day we have iced cocoa or tea, coffee or tea at night. Today they baked us a wonderful apple pie. It's remarkable what the cooks can do with their gasoline field stoves. Fried chicken is a Sunday dish and on a very special occasion, hot biscuits!

We have a little PX of our own next to the office. The PX officer is a nurse who was raised in the Army and is very clever. She has fixed the place up so beautifully that we call it "Lord and Taylor's." A popular feature is a large G.I. can which holds about a hundred pounds of ice, so our soft drinks are cold. The thing we need most, and do not have, is stockings. Unless a supply reaches us soon, we may have to go barelegged. . . .

The American Woman's Club of Bombay sent some money and I have spent 500 rupees to fix up a recreation house. The Services of Supply bought us a very fine piano and we've hired an orchestra and given two dances. Our mess hall is homey. Matting decorated with animal patterns covers most of the wall and in some spots we've hung India prints. The YWCA opened up a very nice little tea room on our "main street" where we can get ice cream, tea and cakes, or sand-wiches. The "Y" also provides a very attractive writing-room and dressing-room. There is dancing at the British Club and American Of-ficers Club.

Captain Dorcas C. Avery

Barbara Drake of Brockton, Massachusetts, graduated from Vassar College in 1931. She joined the Red Cross in August 1943 and was sent to India in January 1944. She traveled on the *Empress of Scotland,* which sailed from Newport News, Virginia, around the Cape of Good Hope to Bombay. While sailing to India, she met Alan F. Hart, a corporal in the Army Air Forces, who was also on his way to southern Asia. They spent the next eleven days together "on the crowded deck" of the ship, where they fell in love. At Bombay, they managed to spend one evening together, and they became engaged. Although Alan Hart was assigned to New Delhi, his job in the Army Air Forces included extensive travel in India. This allowed the couple the opportunity to be together from time to time. During her duty in India, Barbara Drake served as a Red Cross club worker in Calcutta. Her letters to Alan Hart detail a wonderful love story—even in the trying conditions of wartime India.

[Bombay, India
Dear Alan,                                    Early February, 1944]

. . . Why don't we meet at Green's, which is across the street from the Taj Mahal [Hotel], in the lobby at 3:30 p.m. If anything happens that we miss each other, leave a message for me at the Taj Mahal in the box for Room 545 telling me when and where I can meet you. Keep your fingers crossed that we will still be here.

Hastily, Barbara

[On the way to Calcutta]
Alan, my darling,                                    February 14, 1944

It all seemed to end so abruptly the other night that I still cannot believe I won't see you again just any minute. I really hoped that you would be coming with us. I simply floated up to bed the other night, not on brandy I assure you but rather on the intoxication of the night. I don't remember when I have ever been happier. I really worried a little about your getting back to camp as you were feeling almost too high to care whether you made it or not. You were really darling sitting there in the moonlight, telling me things that you had apparently never dared say before.

I rather hoped you might get off the next day too as we did not leave until later than originally planned. . . . The trip through the countryside, much like some of our West, is fascinating although a bit tedious as something falls apart, or the engine has to rest every few minutes. . . . Actually we are fed only twice a day, the dinner after dark so that we can't identify what we are eating. Just as well, because kites [a scavenger bird] dive-bomb us for food in daylight. The children are even filthier and more pathetic looking than those we saw and the poor skinny dogs make me want to weep. . . .

All my love, Barbara

[Calcutta, India]
Alan, My Darling,                                    February 19, 1944

. . . As you can guess, I am most anxious to hear from you, partly to be sure it is all true about us and not just a lovely dream. I wrote your mother a note yesterday telling her that you were well and happy and a little about our adventures, but naturally nothing about our plans. Will you let me know if you are telling your family or do you think it better to wait? I have said nothing to my mother except that we were friends and about the things we had done together.

I am really terribly happy about everything and I certainly hope that you feel the same way. I am so afraid sometimes that it might have been the moon and the brandy and that you may regret having committed yourself. I do so hope not. I have taken your ring to be made smaller so that I can wear it.

Every day I feel surer that I really love you and that it was not just propinquity and the romantic environment. I have inevitably been out a good bit since we arrived, as we all have, but I find I keep thinking about you and seeing your face—in fact, I have been accused a number of times of being absent-minded. . . .

One evening I spent with Mary Jane, who was my best friend . . . and who is working here. We had a dinner sent to her room so that we could catch up on the news without interruption, because one simply cannot get through the hordes of men, British and American, who haunt the lobby and lounge of her hotel unless one rushes through madly as if in desperate need for a little girl's room, or something equally urgent. . . . We were required to go to a tea for some generals, followed by dinner with some brass hats from our ship. Later I was much excited at an enlisted men's dance where I met a boy I know from New Britain (Connecticut) and so it goes at the moment, 12:45, I am tired but I think I'll gradually master the art of turning down dates. . . .

All my love, dear, Barbara


                                                    [Calcutta, India]
Alan Darling,                                       March 1, 1944
I was so thrilled to get my first letter from you yesterday, written on Valentine's Day—do you remember? I do hope that by now you have some word from me as I know how impatient I was.

I am still finding this city fascinating and cannot understand how anyone could be bored. The city is so teeming with activity that it merely seems a matter of having enough eyes in one's head. . . . My dressmaker, Zon, is helping me with Hindustani. . . . [but] I do not feel I am progressing very fast with the language, although I can give directions to cab drivers, know how to order the proper kind of water—bath, drinking, or ice. . . .

On the Red Cross city tours, we visited palaces, temples, and some not so pleasant places. I still cannot respond to the ornateness of Oriental art the way I do to the simpler Greek, early American, or modernistic—too much use of color and confusing detail so that the symmetry of the whole is very difficult to discern. . . .

I am going out into the wilds as I hoped—a rubber boot area, so if you move east, we should run into each other. . . .

All my love, Barbara

Even though Barbara Drake was very much in love, she had, of course, come to India to work for the Red Cross. When writing to Alan Hart on March 29, 1944, she remarked, "Much as I long for your arms around me, we have both known from the beginning that we have a job to do." In this same letter, she also said, "Yesterday I spent time at a nearby canteen to see how it operated: the kind of menus, problems, etc. The day before we had a chat with the soldier committee and again visited the hospital wards where we chat or write letters for the patients." Two days later, she provided a further description of her work.

[Calcutta, India]

Alan, my darling,                                     April 1, 1944

. . . Last night three of the Red Cross girls came in from Burma to do some errands; their life sounds more rugged and exciting than ours. At a party on the base that night we were reprimanded by the C.O. for making too much noise. (But he never really wanted Red Cross girls on his base.) . . .

The first of the week we start painting, carpentry and whitewashing. Your future wife is getting some domestic experience as she had to design and make a sample of a curtain for the durzi to follow. The canteen curtains are bright red with a band of white and one of black on the curtain itself and on the valances. Now I am working on a stylized design of an airplane to applique on the light blue material on our maroon curtains for the little bamboo room. I never made a curtain in my life but I seem to be the only one of the four who can sew at all, and I am finding it fun. . . .

The four of us on this assignment are M., whom you remember meeting in Bombay, J. and V. M. is not going to be too easy to work with as she is a dictator, quite tactless, and obviously because of her physical handicaps, she will never be as popular as the rest of us, and that fact alone will occasionally cause friction. However, the other three of us are so very fond of each other that we'll stick together and be able to handle any situation. Incidentally, M. is very capable, knows what she wants, and stoops to nothing to get it although her methods are much too direct to suit me and much too lacking in feminine charm. J. is my pet really, an Irish girl from Indianapolis, who plays the harp and piano and is a very intelligent, interesting, mature girl. We can talk about life and literature indefinitely. Her husband, a sculptor, is stationed in Italy. V. is a very refreshing, pleasant, friendly girl—absolutely the best type of American womanhood with very high

ideals and an excellent sense of humor. We three shall always cooper-
ate well and trust each other completely. . . .

All my love, Barb


                                                    [Calcutta, India]
My darling,                                         April 18, 1944

You should not have an inferior complex about being an enlisted man
as often the enlisted men do more important work than the officers.
In fact, we find the sergeants, rather than the captains, get things done
for us. We have various enlisted men assigned to us. One who helped
me with our curtain rods the other day said his wife would be amazed
if she could see him doing such a thing, not just willingly, but eagerly.

The days are really fun when M. does other business as her presence somehow spoils the casual atmosphere which we three have deliberately fostered. She is much too business-like and does not seem
to realize how much that we accomplish depends on our working right
along with the men and providing a little incidental entertainment to
make the work palatable. . . . Perhaps cruelly, we avoid going out
with her socially as much as possible as she will engage in conversation
about grease pits, sidewalls, roofs, or cement on little or no encourage-
ment and we do not care to eat, drink and then sleep on the physical
details of the club.

Another beautiful clear day for us. Last night we went out on a
picnic. . . . Although there was no moon, the night was lovely with
millions of bright stars, fireflies everywhere, lights from native fires
reflected in the river, and the sound of drums beating in the distance.
One of those exotic, breath-taking nights that I suspect only the East
can produce because beneath the apparent calm lies an exciting sense
of terror produced by wild animals' eyes, and noises as well as the
delightful uncertainty about the temper of the natives. Actually, this is
all pure imagination as we were indulging in the very American activities of drinking beer and eating fried chicken, but my mind kept floating off in this mysterious atmosphere.

Love, Barbara


                                                    [Calcutta, India]
Dear Alan,                                          August 30, 1944

No, our adventures are not over yet although we hope it is three times
and out. Monday we nearly burned our club down, but so many GI's
appeared from everywhere that the fire was quickly brought under
control. Evidently one of the servants was putting gasoline into a hot

field range; anyway there was a sudden spurt of flame and the back screen, the Hessian cloth on the back wall, the tarpaulin and the tar on the roof were on fire. I swear Ali Akbar threw some of our lemonade before the boys' arrival with a fire extinguisher. Meantime Ginny and I had been carrying everything that was not nailed down out of the lounge.

It just so happened that our new Club Director arrived that very day and hence her first view of the club was the mess. Her name is Mary Anderson, from Nashville, Tennessee, aged 39, small, dark-haired, blue-eyed and very attractive. She spent 18 months in a Red Cross club in Belfast, Ireland. We all talked very frankly today so that there could be no undercurrents. The suggestion from headquarters was that we three be split up or if she preferred, we could all be transferred, but she would rather we all stayed awhile—the happiest solution. . . .

Love, Barb

[Calcutta, India]
Dear Alan: November 2, 1944

. . . Our Hallowe'en party at the club was quite a success—pumpkins in traditional fashion, witches, cats, and ghosts pasted on the walls, games and stunts. The best was a relay race with the boys trying to put on the dhoti, the Indian version of male pants. Ahmedulla, our cook, was the judge and I thought he would have hysterics at the various draped effects achieved. He was not in the least insulted and in the end demonstrated the correct way.

Love, Barb

[Calcutta, India]
Dear Alan: November 6, 1944

. . . Last night we were disturbed by what was apparently a jackal wandering about in our back storeroom and on the back porch. I suppose he would not really harm us but I don't like hearing anything prowl.

Today a girl came through from China on her way home. . . . I think she looks terrible. She has lost 20 pounds and looks a dozen years older with deep lines on her face. Life has been very rugged for her—the food was terrible and she was sick a great deal. Perhaps it is just as well that you and I have kept out of China after all. . . .

Your Barb

[Calcutta, India
Dear Alan:                                    December 15, 1944]
I am still anxiously awaiting word from you as to whether you are
coming. I almost feel not hearing from you is a favorable sign. . . .

We have started making tiny red stockings to hang about the fireplace at the club. We'll put you to work if you come.

Love, Barb


Shortly after Barbara Drake wrote this letter, Alan Hart arrived in
Calcutta for a twenty-day furlough, and the couple formally announced their engagement. On August 25, 1945, they were married in
Bangalore, India, where Alan Hart was then stationed. They honey-
mooned in the British Hill Station at Ootacamund. At the end of the
war, Barbara and Alan Hart returned to United States where Alan
resumed his pre-war position in the Treasury Department of the State
of New Jersey. Barbara Hart taught English at Rider College and,
later, at Hopewill Valley High School. Alan Hart died in 1980. Barbara Hart, who lives in Pennington, New Jersey, has traveled widely
since retiring from her teaching position.

CHINA WAS EVEN MORE remote than India and presented immense
challenges to Red Cross workers. One of the first Red Cross women
to arrive in China was Rita Pilkey of Dallas, Texas. Pilkey, a 1929
graduate of North Texas State College, joined the Red Cross in the
autumn of 1943. After completing Red Cross training in Washington,
D.C., she spent six weeks in New York City, where she ran "The
Dugout," a facility that provided cafeteria-style meals for service personnel.

Late in 1943, Pilkey was sent to the CBI theater and assigned to
China's Yunan Province. Getting to China was not easy. She took a
train across the United States from Washington, D.C., to Long Beach,
California. From there, she traveled in a merchant ship to Hobart,
Tasmania. The ship then went to Perth, Australia, and from there to
Bombay, India. After disembarking in Bombay, Pilkey traveled by
train across India to Red Cross CBI headquarters in Calcutta. Fourteen Red Cross women out of a pool of 120 were then chosen for duty
in China. Pilkey and her Red Cross colleagues flew over "The Hump,"
landing in Kunming in late January 1944. At that time, there were

fewer than forty Red Cross personnel working in all of China. From Kunming, she traveled by motor vehicle to her duty station about fifteen miles away at the Army's Y-Force Field Artillery Training Center, where she served as Field Director and, after a promotion, Club Director for the men stationed at the Center. Her letters to her parents provide a close-up and discerning look at the life of a Red Cross club worker assigned to remote outposts in China.[14]

Dear Mama and Papa,

[Somewhere in India
January 12, 1944]

. . . We have all just received our assignment to the part of the country we'll be in and I am thrilled to pieces over mine. . . . We have had lots of fun here, and have done our part too. One day we rode out on Army trucks to a camp that had just been put up and took our mess kits to eat supper with them. Then we made doughnuts out there in the open and cooked them and coffee and served to the whole camp. We really cooked lots of doughnuts. . . . We rode in open trucks and the roads are like West Texas, they are so sandy so we were really dirty. . . .

Rita

Dear Mama and Papa,

[Somewhere in China]
February 9, 1944

. . . I guess you have received my letter saying that I am in China now, and I guess I'm settled now for at least 6 months. . . . I'm sorry I can't tell you where in China I am, but you'll just have to look at a map and guess. Instead of having big clubs in the cities like the U.S.O. at home, there are clubs on each post with two girls to man it. We are in a wonderful set up—with the Field Artillery Training Center with a General [Jerome J.] Waters, an ex A & M from Texas. They are training the Chinese soldiers here.

The climate is wonderful, and the evening looks like Colorado. We have a little house all our own with a bedroom each, a bath between, and a little sitting room. We have a Chinese houseboy who cleans, builds a fire, (charcoal stove), makes the beds, and does anything we want him to. He doesn't speak English and we're not so good at Chinese yet, but we get quite a kick out of trying to make him understand. . . .

14. The letters of Rita Pilkey are located in the Rita Pilkey Collection, Blagg-Huey Library, Texas Woman's University, Denton, Texas. Hisgen, "American Red Cross in China, Burma, and India, 1942–1946," 119.

We run the club for the boys, going up there every evening and most afternoons, planning parties and activities. We've just had a birthday party, we're finishing off a pinochle tournament, and are starting to work on a minstrel. The last of the month we are having a Barn party and inviting all the Red Cross girls in the area. Thursday and Sat. night we are going to dances in the other hostels. . . .

I'll write again soon, Rita

[Somewhere in China]

Dear Mama and Papa,                    February 26, 1944

. . . We have been out here at our assignment long enough to get started and get settled and get in the swing of things and it is loads of fun. Work too, but I guess I'm used to that. . . . Tonight [co-worker] Lucille [B. Young of Detroit, Michigan] and I have the evening off because our boys are having a meeting. It is our first evening off since we have been here and we are really enjoying it. . . . This morning we went to town [Kunming] in a jeep, went to Special Services to remind them that we need a guitar, picked up pictures at the photo lab, did a few more errands, and this afternoon we covered the bulletin board at the club and put up all the pictures.

The movie star, Paulette Goddard, was here and with three men, gave a performance. The General had a dinner for her beforehand, and then we had an enlisted man's reception in the club after the show. . . . I think I told you what a cute little house we have and how nice it is fixed up. The nicest in China, and we have visited all the other hostels. Our General flew to India and brought our drapes. The gardener brings us flowers once a week when he goes to town, and right now I have some in an old shell shined up for a vase. The fellows have made us an ironing board and you should see our M1 (Model One) pantie hangar for drying our clothes on. It even has rope tassels on it. . . .

We had a wonderful barn party Saturday night. Our club was all decorated with hay and bridles, etc. and one of the boys played my accordion, another a fiddle and one a banjo and one of the boys called. They all had a wonderful time. Right now we are working on a minstrel show and it is lots of fun. . . .

Lots of love, Rita

Due to the dangerous conditions that prevailed throughout much of China, Red Cross workers were instructed in the use of firearms. In the next letter, Rita Pilkey provides a brief account of this training.

[Somewhere in China
March 8, 1944]

Dear Folks:

. . . Yesterday Lucille and I fired all kinds of guns in a small arms class. Everyone on our post is taking a 3 day review and so we got in. . . . This morning we got up early and went out on a problem with a group. They packed the mules with guns etc. and all the Chinese soldiers wore some shrubbery and away we went to the hills to set up the 75's. It was very interesting. . . . We're going to learn communications next, learn to work the walkie-talkie etc. . . .

Tonight is a big night. We are going to a dinner dance given by General [Claire] Chennault [Allied Air Commander in China and former Commander of the "Flying Tigers"] in honor of the first anniversary of the formation of the 14th Air Force. We got a very formal invitation, and we are going with our General. There are so few American Girls here that we all get invited to everything. . . .

Lots of Love, Rita

[Somewhere in China]
May 25, 1944

Dear Folks,

. . . I feel terrible for not writing sooner, but time has been flying by in some of these places. I think I told you about working on a Minstrel Show. Well, it went over with such a bang that we have been taking it around to the nearby bases and having a whale of a good time. . . .

We had something new about a week ago, a jingbow (air raid) about 4 a.m. in the morning and we had to get up, don slacks, helmets, etc., and go up on the hill to the slit trenches. Needless to say, Lucille and I were the first ones up there. After about 30 minutes with nothing happening, it was all clear and we went back down to bed. . . .

Love, Rita

Given the harsh realities of Red Cross work in China, the ARC recognized the need to provide rest camps for its staff.[15] In the next letter, Rita Pilkey describes her visit to a nearby rest camp on Kunming Lake.

[Somewhere in China]
August 6, 1944

Dear Folks,

. . . I am on a week's vacation at Rest Camp. I think I told you about

---

15. Hisgen, "American Red Cross in China, Burma, and India, 1942–1946," 120.

driving down for the day one time, well, every Red Cross girl who wishes to, may spend a week here, and I am here this week and Lucille next week. It is a place designed for service men to have a week (or more if they need it) of change and rest. In other places, it is just for air corps, but in China it is for everyone because there is no place to go on a leave. It is in a beautiful lake away up in the mountains (over 8,000 ft.) and there are motor boats, sail boats, lots of those rubber life rafts to play in, good swimming, 2 tennis courts, volley ball. It is a beautiful place and the best food in China. . . . There are 2 nurses here and no other Red Cross girls, but there are about 20 fellows from our camp, besides all the others. So I'm sure it will be loads of fun. I've just been here a day, but I breakfasted with a colonel, lunched with a few sgts., and had supper with a captain and a major. I've been swimming twice, motor boating, played basket ball, been in a rubber boat, played ping pong, took a nap, am writing you and then I'm going to a movie (probably years old, but a movie nevertheless). . . .
[Love, Rita]


                                                   [Somewhere in China
Dear Folks,                                         August 1944]
. . . I almost forgot to tell you that I got a promotion from a Program Director to a Club Director. It will mean an increase in salary, but I'm not sure how much, and it also means that the next big club that is opened in China, I'll have charge. There is really supposed to be a man at the head and then a woman, a program director and 2 staff assistants, but we don't have men in the clubs here. Girls are so scarce that they don't like to take up tonnage sending men over. The night I was promoted, we had a celebration at our house after club hours and had lots of fun. . . .
All my love, Rita


In the next letter, Pilkey wrote that she had received a new assignment. She would now be stationed at the Army Air Forces base at Luliang, where the "Flying Tigers" had once been stationed.


                                                   [Somewhere in China]
Dear Mama and Papa,                                 November 13, 1944
. . . I am going soon to open a new club at a new base—the club will be in a tent. I love it here, but I am looking forward to tackling the new job. I am going to fly down tomorrow with my supervisor to look

over the situation. I'll tell you more about it when I know more. Write
to me soon,

Lovingly, Rita


                                                    [Somewhere in China]
Dear Mama and Papa,                                  November 19, 1944

. . . I have been very busy this last week working on the new club I
am about to open. . . . I flew down with the supervisor for a couple
of days, and it is just like the worst part of West Texas with nothing
but dust. It will be quite a change from the place where there are
flowers all over and the side of the hill terraced, but there are a lot of
fellows there and they have nothing so I'll be very glad if I can bring
them a little happiness by manning a club there. The girl who is going
with me seems to be very sweet, and we have been working requisi-
tions etc. to get going. We are going down to see how near things are
ready and we are going, so if I don't write for a week or two, you'll
know that I am head over heels in work and I'll write when I can.
This will be the first tent club in China. . . .

All my love, Rita


In a long retrospective letter written to her parents in March 1945,
Pilkey reported that "by hard work, we were able to open two of our
tents for an Open House on Christmas. . . . It was a big success and
the boys all seemed to appreciate all our efforts so much that we decided to keep those two tents open while we continued to work on
the rest of the club." In the same letter, she described the grand open-
ing of the Canvas Cover Club on January 30, 1945, complete with a
big tent for a Snack Bar and smaller tents for a lounge, library, game
room, and store room. She also noted that "we have terrible dust
storms and the wind blows a terrific gale almost constantly, and often
the tents seem as tho' they are going to take off like a balloon."

Pilkey underscored the challenges she faced at her new assignment
in Luliang in her ARC report dated March 31, 1945: "We have had
three consecutive days of sunshine with no excess wind or dust or
rain, and no calamities such as fires, trouble, or running out of something that we can't do without. This is definitely a deviation from the
normal trend."

In the next letter to her parents, Pilkey offered a description of a
more permanent structure under construction for the Red Cross at
Luliang.

[Somewhere in China
Dear Mama,                                      March 31, 1945]

. . . You should see our new building. It is almost finished and the
most beautiful building I've seen in China, and what a wonderful
satisfaction after the struggle I've gone through here to get everything.
Our old tents have had a lot of wear, but they have meant a lot to
the boys. The place is always crowded and we have wonderful
doughnuts.

As you can see by my letter [in which she enclosed a copy of her
monthly report to her headquarters], I've had malaria. It's the first
time I've been sick since I've been overseas and it really made me
mad. I got off light. The dr. caught it right at the beginning and I am
taking Atabrine for a month so he doesn't think there will be any
reoccurrence. We were very lucky to have moved out of tents into our
new quarters before I got sick, and we're right next door to the hospital so I stayed at home and they came over to take care of me. Atabrine gives you a yellow tinge to your complexion which gives me a
sort of olive complexion, quite becoming if I don't become more
yellow. . . .

With love, Rita

 

[Somewhere in China]
Dear Mama and Papa,                             April 20, 1945

. . . As I told you in the V-mail, I am spending a week at Rest Camp,
taking a week of much needed rest. It has been pretty tough going
on this new assignment and I have been operating the tent club with
very little help and with a crew of Chinese waiters, cooks, coolies,
that don't speak English. I could drive myself all day and evening
by then I'd be so tired that I just couldn't do any more, hence no
letters.

Everything is changed now tho'. I have a good staff, a missionary
who speaks Chinese fluently, and I am having a good rest, so I'll have
new inspiration and vigor to open our new club.

The penalty of postponing my rotation because of an unauthorized
plane trip [to obtain supplies, including a piano, for her new club] has
been changed due to my past service. However, I won't be home this
summer because my new club will just about be finished and I think
I'd better get it opened and not leave a half done job. . . .

I think I told you I received nine packages in two days . . . and
they came at such an opportune time. They had just moved the tents
and all to the other side of the field; we were trying to settle our things
again, the food in the mess hall was terrible, my penalty was hanging
fire with everyone making a to do over it, and the Supervisor came

down and settled it, and we all celebrated and fixed a good supper with the food I had received. We had the best time we had had for a long time. . . .

Miss. S., the missionary is a nice old woman but not too much help other than being able to speak Chinese quite well, but she is going to supervise the boys fixing the garden. . . . You would die laughing at us here, the Chinese steal anything they can get their hands on, and when something disappears, Miss S. will pray for their poor characters while I go over and threaten them with their lives and get the stuff back. I don't speak Chinese so well but I have been operating the club and canteen with 6 waiters, 3 pantry boys, 2 cooks, and 4 coolies without an interpreter and believe me, they understand me. . . .

Love and Kisses, Rita


                                              [Somewhere in China]

Dear Mama and Papa:                         July 4, 1945

. . . Well, we have said goodbye to our Canvas Cover Club, taken the tents down, and are working full time on trying to open our new club. It is going to be the prettiest one in China, for sure. Our furniture has started arriving by truck, and the part that is being made for here is coming in. I've been up to the city [Kunming] to go through our Regional Warehouse for our curtain material and are expecting it at any moment. Things are really shaping up. . . .

I had a very interesting trip last week. The supervisor was here and I went with her and the Field Director in a jeep to a new base where they wanted to start a Red Cross Club. It was a beautiful drive through the mountains and you know here in China, when you drive any place away from the base you are on, you always take a gun because of bandits. That always makes a trip more interesting. . . .

It seems a shame to work from nothing and finally to get the best club in China, and then leave it for some one else to run, but what of it, I'm coming home for awhile. . . .

I'm getting to be quite valuable. My hair is turning to silver and my skin turning to gold (from Atabrine.)

Take Good Care of Yourself, Loving You, Rita


Rita Pilkey was on her way to the United States for a home leave when the war ended. She did not return to China. She received the American Red Cross Certificate of Outstanding Service for her overseas work. After the war, she earned a master's degree in physical education from Columbia University. She then returned to Texas, where she taught physical education at the University of North Texas until her retirement in 1974, when she was appointed professor emerita. She travels extensively and lives in Denton, Texas.

LORRAINE "BUBBIE" TURNBULL was one of just one thousand Women Marines who, during the early months of 1945, were selected for overseas duty in Hawaii.[16] In February 1945, she was assigned to the Marine Corps Air Station at Ewa, where her skills as a flight mechanic were put to good use. The letters she wrote to her family from her duty station at Ewa contain enthusiastic accounts of her work as a flight mechanic. This selection concludes with Turnbull's victory letter. (For other letters by Turnbull, see pages 51–52.)

[Territory of Hawaii]

Dearest Mom,                                    February 20, 1945

. . . At present I'm waiting to go over to the Administration Building to see the Captain. I think they are going to put the mechs out on the line so I'll be one happy gal because that is exactly what I want to do. We hear and see the planes all day and night. They take off and come over our barracks with a roar, and I love that roar.

They keep us busy G.I.ing this place, cleaning up the laundry, etc. There is a coral dust which collects real fast. Our barracks are splendid as I said before. Big roomy lockers. . . . If you forget where you put something, you have to look for an hour for it.

The weather is good here. Just like a balmy spring day at home. The cloud formations are beautiful. Mountains aren't far away, and the green lines of the vegetation and the purplish-red soil is beautiful. . . . Between our barracks we have some beautiful trees, and when we get grass and flowers and shrubs planted our area will really be something to look at. . . .

I'll write to you soon, Mom.

Love, Bubbie

---

16. Pat Meid, "Marine Corps Women's Reserve in World War II" (Washington, D.C.: Historical Branch, U.S. Marine Corps, 1968), 50.

[Territory of Hawaii]
Dearest Mom and Johnny,                              April 5, 1945

. . . The big news—I've started work (2 April) to be exact. And I love it. I am crew chief on an S.N.J. Texan [a single-engine monoplane]. I take care of it; inspect it each day, rev it up, gas and oil it, and keep it in "going" condition always. When a pilot takes my plane, I have to guide the plane out and then bring it in. It's what I've dreamed about ever since I joined the M.C. We go to work at 0730 in the morning to warm-up and check our planes. I haven't decided what to call her yet, but when I do decide, I'll get it painted on. Got any suggestions? . . .

Lovingly, Bubbie

[Territory of Hawaii]
Dearest Mom,                                      August 15, 1945

As we all said yesterday, "this is it." I keep trying to beat into my brain that it's all over. Isn't it wonderful? I'm so thankful that there won't be another invasion and no more lives lost. . . . Last night we went to church at 7:15. Both the Protestants and Catholics had services. We had several hymns, a sermon and then they played taps. We then walked home. Mary got her Lts. jeep and we went riding. The search lights from all the A.A. posts and all the ships in the harbor played all evening. The streams of light which looked like silvery fingers whipped back and forth all evening. . . .

Tomorrow we will go back to work as if nothing had happened. Now just because the fighting is over, you can't expect us home. I don't know whether we will have to finish out our assigned tour of duty. We've heard about ten versions of the point system which is supposed to have been recently concocted. . . . [The War Department constructed a point system of demobilization based on service longevity, overseas duty, combat experience, and parenthood.]

Will write again soon, Bubbie

# Preparing for the Postwar World

PRESIDENT HARRY S TRUMAN'S proclamation of the cessation of hostilities on August 14, 1945, called for a nationwide holiday and a period of thanksgiving. Throughout the United States, millions of Americans triumphantly celebrated the end of the Second World War. Yet for many uniformed personnel, victory was celebrated with restraint, for they, more than most civilians, understood the difficult tasks that lay ahead. Writing from Germany on V-E Day, a Red Cross worker observed in a letter to her mother, "It isn't that we weren't happy that Death's day was done and that home was nearer, but only that we experienced none of the wild joy we had expected." [1]

Combat and killing may have come to an end, but a multitude of tasks still faced uniformed personnel, especially those who were over-seas. Released prisoners of war needed to be restored to health and reunited with loved ones. The survivors of the concentration camp "hell-holes" were in desperate need of food, medicine, and civilized human treatment. The "Displaced Persons" who poured into Western Europe with the end of the Nazi machine also needed the succor that could only be provided by the victors. Nurses and doctors began the work of rehabilitation even as the war was still raging. Within a matter of weeks after the war ended, the United Nations Relief and Rehabili-tation Agency (UNRRA), the first service organization of the United Nations, took on the task of alleviating the great human misery that blighted much of Europe and Asia.

---

1. Dorothy J. McDonald to her mother, May 8, 1945.

Most women who donned uniforms in World War II agreed then, and recall now, that they were significantly changed by their wartime experiences. The challenges presented by their years at war had enabled them to become stronger individuals with a clearer sense of themselves and their capabilities. Yet the postwar world that loomed before United States women presented a picture that was dim and unclear. The incredible devastation of the Second World War, capped with the horrors of Iwo Jima, Okinawa, Dachau, Belsen, Hiroshima, and Nagasaki during the last six months of the fighting, was such as to give serious persons much pause for thought. The future was not known, and the prospects were sobering for many. What was generally understood, however, was that United States women *could* and probably *would* play a much stronger role than ever before in determining what the future would bring.

ANNABEL GOODE of Virden, Illinois, joined the WAVES in June 1943. She received instruction in the operation of the Link Trainer, a flight simulator used to train prospective aviators. In July 1945, she was transferred from Kingsfield, Texas, to Kaneohe Bay in Oahu, Hawaii. Five of the Goode family were in the service during World War II, and the family published a newsletter, *The Goode News,* as a way to pool information from all the fronts and keep up with the whereabouts of each other. The issue of September 4, 1945, carried Annabel Goode's reaction to the news of the end of the war.

                                        Kaneohe Bay, Oahu, T.H.
Dear Family,                            August 14, 1945
This is the day the whole world has been living for. Pau! (the end). I consider it a privilege to have received the news of peace here on the very island where our nation's part in the war began. I was at my Link desk giving a hop when the official message came over the radio this afternoon. Though we have been expecting it hourly since Saturday, there was nevertheless a unique thrill at the moment. A silent surge of happiness seemed to possess each one of us. Our thoughts raced backward to December 7, 1941 and then forward to the civilian future we have hardly dared to contemplate seriously until today. Then, tying our four years of memories into one package with a prayer, each went on with his own occupation and plan for celebration. . . .

Here's hoping you can all be home for Christmas. I've been here for such a short time I rather hope to be among the last to leave. . . . me ke aloha pau ole [everlasting affection.]

Ann

On the other side of the world, the news of the bombs that dropped on Hiroshima on August 6 and Nagasaki on August 9 came as a relief to those who assumed that they might be part of the assault force on the islands of Japan. Many Red Cross workers in Europe had faced the horrors of the Nazi war and were caught up in these revelations. The following letters, written by Red Cross clubmobiler Mary "Chichi" Metcalfe in the late spring and summer of 1945, depict her reaction to the Nazi atrocities as well as her muted response to the end of the war.[2] (For other letters by Metcalfe, see pages 151–155.)

[Somewhere in Germany
April 28, 1945]

Dearest Mommie,

Lea had the day off yesterday and I am off today. How I have enjoyed doing some much needed sewing and washing and listening to the radio. I had almost forgotten what it was like to be by yourself and not bouncing for miles over rough roads, making one batch of coffee after another, and putting out a line of chatter every minute. . . .

You ask if travel over here wasn't terrific for people to be going to the Riviera on leave. The Riviera is a leave area and it is necessary for any of us over here to have leave in order to keep going. . . .

There are beginning to be stories about all the looting that is taking place. However, it makes you mad when you see how well these Germans have been living off of everyone else, and it is time some of it is taken from them. All of these people have had Russians doing all their labor for them. There are several concentration camps near here and the boys have showed us snapshots of the stacks of dead people, nothing but skin and bones, lying there. These boys actually saw these awful sights and they all say it is something you couldn't believe unless you see it. The Army made the Germans come out and dig graves and bury them. It seems that nothing is too bad for people who would just stand by and let something like that go on. . . .

Loads and loads of love, Chichi

2. The letters of Mary Metcalfe are located in the Mary Metcalfe Rexford Letters, Papers of World War II Participants And Contemporaries, Dwight D. Eisenhower Library, Abilene, Kansas.

[Somewhere in Germany]
Dearest Mommie,                                                    May 3, 1945

How wonderful the news has been these last few days. I imagine it is much more exciting getting the news there than it is here, in fact. I wouldn't be surprised if you know the news of the war's end before we do. They certainly can't hold out much longer now the way everything is crumbling. . . .

Loads and Loads of Love, and Happy Mother's Day, Chichi

[Somewhere in Germany]
Dearest Mommie,                                                   May 10, 1945

Well, it's finally all over, over here. It must have been quite exciting at home on V-E day. It was just like any other day over here. Everyone had expected that V-E day might be announced every day for a week or more so it was not a great surprise. Also it is not as though it were all over because our boys are wondering what next for them. Will it be Army of Occupation or the Pacific? . . .

Lots and lots of love, Chichi

Metcalfe's weekly letter to her mother at the time of the news of the Japanese surrender made only a brief reference to the end of the war. For Mary Metcalfe, as for most of her colleagues, it was just another day in Germany.

Hofgeismar, Germany
Dearest Mommie,                                                August 17, 1945

. . . The roof of poor old Abe [her clubmobile was named "Abraham Lincoln"] leaks so badly now that when we got out to serve the other day having driven an hour and a half in the pouring rain to our first stop, I got in the back to get out the doughnuts which we were going to serve in the chow line and found that every box was soaked through. 1500 doughnuts ruined. What a sight those awful soggy doughnuts were. It's bad enough to look at them when they are in good condition.

The old bus is going into ordnance on Mond. for a complete redoing—painting, inside and out, all sorts of gadgets fixed, roof repaired and motor overhauled. . . . You must be beside yourself with joy at the news of the past few days. . . .

Loads and loads of love, and Happy Birthday, Chichi

At the end of the war, Army nurses, such as June Wandrey, found themselves working to save the lives of the survivors of concentration camps. The next letter describes her experiences with hospitalized victims from Dachau.[3] (For other letters by Wandrey, see pages 125–129.)

Allach, Germany<br>June 4, 1945

Dearest Family,

I'm on night duty with a hundred corpse-like patients, wrecks of humanity . . . macerated skin drawn over their bones, eyes sunken in wide sockets, hair shaved off. Mostly Jewish, these tortured souls hardly resemble humans. Their bodies are riddled with diseases. Many have tuberculosis, typhus, enterocolitis, (constant diarrhea) and huge bed sores.

Many cough all night long, as their lungs are in such terrible condition. They break out in great beads of perspiration. Then there is the room of those who are incontinent and irrational. It sounds like the construction crew for the Tower of Babel. . . . Poles, Czechs, Russians, Slavs, Bulgarians, Dutch, Hungarians, Germans. What makes it so difficult is that I understand only a few words. Their gratitude tears at my heart when I do something to make them more comfortable or give them a little food or smile at them. . . .

The odor from the lack of sanitation over the years makes the whole place smell like rotten, rotten sewage. We wear masks constantly although they don't keep out the stench. There are commodes in the middle of the room. Patients wear just pajama shirts as they can't get the bottoms down fast enough to use the commodes. God, where are you?

Making rounds by flashlight is an eerie sensation. I'll hear calloused footsteps shuffling behind me and turn in time to see four semi-nude skeletons gliding toward the commodes. God, where were you?

You have to gently shake some of the patients to see if they are still alive. Their breathing is so shallow, pulse debatable. Many die in their sleep. I carry their bodies back to a storage room, they are very light, just the weight of their demineralized bones. Each time, I breathe a wee prayer for them. God, are you there?

. . . Our men sprayed the camp area to kill the insects that carried many of the diseases. We were told that the SS guards who controlled the camp used to bring a small pan of food into the ward, and throw

---

3. The June 4, 1945, letter of June Wandrey appears in June Wandrey, *Bedpan Commando: The Story of a Combat Nurse During World War II* (Elmore, Ohio: Elmore Publishing Company, 1989), 204–6.

it on the floor. When the stronger patients scrambled for it, like starving beasts, they were lashed with a long whip. It's a corner of hell. Too shocked and tired to write anymore.

Love, June


An estimated fifty million individuals, both civilian and military, died as a result of World War II. In the next selection, Erna Elizabeth Maas, an Army nurse stationed in Germany, chronicles how she personally coped with the death of one of the war's casualties. Maas, a graduate of the East Orange (New Jersey) General Hospital School of Nursing, joined the Army Nurse Corps in December 1943. After serving with the Eighty-fifth Evacuation Hospital at Camp Swift, Texas, she was sent with the Eighty-fifth to Europe in January 1945, where she served in both France and Germany. She returned to the United States in January 1946 and was discharged. She now lives in Prescott, Arizona. Shortly after Erna Maas arrived in Europe, her Army boyfriend, who was stationed in Germany, was in a serious automobile accident and died several days later. Yet Erna Maas was philosophical about this tragedy, as she recognized that "everything happens for a reason and our good."


                                                            [Germany]
Dear Folks,                                               May 26, 1945

Have been waiting for mail from you but none tonight again. Letter writing has become an ordeal lately. Haven't written to anyone but you. . . . Major Gillman took Ruth and I to Liege, Belgium for the day. He promised to take me to Doc's grave so we stopped at Aachen at the hospital and found out where they put him. He is about twelve miles from there, just outside Maastricht, Holland in a U.S. Military Cemetery, with thousands of others. There is just a small white cross with his dog tag in the center. They haven't fixed it yet. Just a big pile of dirt in front of the cross. I hope to go back some day and put some kind of flowering bush on it. I'm glad he's in Holland. The surrounding country is beautiful. . . . The most beautiful part of the trip was seeing field after field of poppies. Stopped to pick some but didn't because we were afraid of mines. They are absolutely gorgeous. In some fields they were mixed with blue bachelor's buttons. . . .

In my wildest dreams I never thought I'd be seeing as much of the world as I am. What do you think of your daughter spending seven days on the Riviera [where she was about to go on leave]? . . .

Wouldn't have missed this experience for the world, but often think of you folks and that nice room I had on Norman Street. Hope you are all well and not worrying about me. The going has been a little rough at times but good for me. I'm growing up and much more tolerant and understanding than I used to be. So no matter how my letters sound some times, know that I am enjoying it and profiting too. Everything happens for a reason and our good. That's the only way to look at it. . . .

Love to All, Erna

 

Ellwangen, Germany

Dear Mom and Pop,                  V-J Day

Well, it's all over. I can imagine how happy everyone at home must be. I'd love to be there celebrating too. We're having a party here tonight. I don't often go to their parties, but think I'll go over for a little while tonight. Have been making postwar plans already. Thinking of how much fun it will be to buy civilian clothes and dress up again. Actually though, I think I will be lucky if I'm out a year from now. The war isn't over for us. Oh, well, it's something to look forward to any way. Everyone is thrilled, but it has been very quiet. Just like any other day. . . .

Boy, Japan didn't last long after Germany went kaput. Our prayers were answered much sooner than anyone expected. I do hope, though, that they throw away the recipe for the atomic bomb. We may live to regret the day that it was dreamed up. . . .

Love to all, Erna

 

Ellwangen, Germany

Dear Mom and Pop,           August 30, 1945

. . . Gee, it will be wonderful when everyone gets home again. I can't even imagine it. I'll probably be the last one so am looking forward to a nice reception committee. All the thrill and excitement we missed on V-E and V-J days will be wrapped up in that one. Almost seems too good to come true.

Well, this is my 5th nite—9 more to go. They are just bringing a boy up from the operating room, Appendectomy. Next room we have a 27 year old captain, dying from a skull fracture. Jeep accident. He's married. His case is very similar to Doc's. I used to wish I had been able to stay with Doc until the end, but changed my mind while sitting with this boy last nite. Saw him once and am glad but don't think I could have taken the end. This boy is more fortunate.—He's going much faster. Just imagine his wife at home. Thankful the war is over and expecting him home soon. A friend of his told us he came in a few days after D-Day. Received a bad chest wound, but was pulled through. Went on up through France, fought through the Battle of the Bulge. Received several shrapnel wounds but made it ok and now after the war is over, this. It would almost be more justified if he had died a battle casualty. Just can't seem to forget Doc. There are so many reminders. Still have his clothes and camera all packed to send home, but when I do, I'll have to write to his mother, and somehow just haven't been able to do it. Thought of her on V-J day and know how she must have felt. Took all the joy out of victory. Just thanked God that no more would have to give their lives. We have been very lucky in our family and circle of friends.

The patient passed on at 4:18 this morning. His wife will be receiving that heartbreaking War Department telegram in a few days. Hope there aren't any children. People who say nurses and doctors are hardened to pain and death don't know what they're talking about. . . .

Love to all, Erna

The responsibility of tending to the debilitated civilian populations of Europe and Asia was soon handed over to a multi-national medical team who reported to the United Nations Relief and Rehabilitation Agency. A letter written by an American UNRRA nurse in Europe describing her work appeared in the September 1945 issue of the *American Journal of Nursing.*[4]

[Halle, Germany<br>Early Summer, 1945]

[Dear Editor of the *American Journal of Nursing:*]

We walked into a prize assignment. The Military Government in control made us responsible for all DP's [Displaced Persons] in the area of Halle: twenty-four camps and nearly forty small groups of eight to one hundred people scattered around the city.

We are trying to centralize them into eight large camps of from three hundred to eight thousand. Another UNRRA team came in the other day, and they have taken over the largest. . . .

We have sent home about 12,000 DP's. Dusting them before they leave is a job, particularly when we have short notice. A trainload of

4. Reprinted with permission from the *American Journal of Nursing* 45 (September 1945): 954–55. Copyright 1945 The American Journal of Nursing Company.

Dutch left the other day on twelve hours' warning. The last of the 1,700 caught the flit gun as the train was pulling out, but they got their share of DDT.

Ann F. Matthews, R.N.

Just as the task of ameliorating the widespread human devastation in Europe was of great importance, so, too, was it imperative to come to the aid of the millions of war victims in the Pacific. Following the liberation of the United States territory of Guam in July 1944, Navy nurse Lillian Schoonover of Knoxville, Pennsylvania, was assigned to U.S. Government Hospital No. 203 in Guam, where she helped restore the island's population to good health. One of her tasks on Guam was to reopen a nurse's training school for the native women that Navy nurses had operated before Guam fell to the Japanese in 1941. After her tour on Guam, Lillian Schoonover was transferred to the Philadelphia Naval Hospital. She retired from the Navy Nurse Corps in 1968, and today she lives in her hometown of Knoxville.

[Guam]
Dear Folks, June 27, 1945

. . . All things point to our having a great experience out here. It seems to me that my job is pretty big. . . .

I love you and several times a day think of being back home again. This is selfish—very selfish in the midst of all this need. You can not imagine. Seeing is believing, and they say it is much improved over six months ago.

I am in operating room, very rusty on technique but hope to do some scrubbing myself one of these days and of course, I still retain the nurse's conception of sterile technique and *bedside* surgical nursing. The native nurse in charge of O.R. is good. . . .

June 28, 10:10 p.m.

It is 11 a.m. tomorrow at home and I suppose you are thinking what to have for lunch. . . . Today has been a busy one and the *work* I have confronting me is tremendous. I am going to start staying in at our hosp. after hours in order to *begin* to do something. Daytimes, I am too busy. In the morning in Operating Room and in the afternoon in the surgical wards. . . .

All my love, Lillian

Somewhere in Guam
Good Morning, Folks!                                    July 30, 1945

. . . I plan the menu for the hospital, making adjustments from the regular diet for special ones such as softs, higher protein, low salt, low carbohydrates, etc. Then I see that the man in charge of the galley keeps it going and the workers as happy as possible, check on the native dietician's work. These last two duties are not much trouble to me as the galley watch and dietician are so good. I'm supposed to be supervising Men's Medical and Surgical Wards, but haven't had time to get over there much. . . .

All my love to you, Lillian


Agana, Guam
Dearest Mom and Dad,                              September 13, 1945

. . . It rains a great deal here now. Sometimes the rain beats the hut roof so hard one has to shout to be heard. The huts are made of this corrugated tin, you know, and the rain on the roof situation creates a great din. We have heavy Navy issue shoes to wear in this weather. . . . We also have ponchos and rain helmets. Did I tell you about the ponchos? They are camouflaged canvas. . . . When they are donned, they assume the shape of the body to some degree and are the most effective rain clothes I ever saw. . . .

Well, at last I'm teaching school. I have 11 native boys who are to be corpsmen. . . . They started class on Monday and do as well as they can considering language, ability, and what the past four years have done to them. I am much happier teaching than not and the slowness with which we proceed is the least of my worries. . . .

Love, Lillian

P.S. The corpsmen's classes are *extra* and not part of our school which still hasn't formally started.


Agana, Guam
Dearest Mom and Dad,                              September 8, 1945

The lid is off the censorship, so here goes my first unexamined letter to you. . . . The ruins of Guam's capital city are where we are located. It is pronounced, Agana, accent on the middle syllable. Great changes have taken place ever since we came, as far as cleaning up wreckage and debris goes. It hardly looks like the same place and was somewhat on the dismal side to be plunked down in the middle of a few Quonset huts with nothing but shells of buildings and piles of wreckage around. . . .

All my love, Lillian

Agana, Guam
September 30, 1945

Dear Neighbors,

. . . Our work here is to start a school of nursing for native women. It will take a long time to get it as we want it. Most of us will not see the finish of the school for it is reasonable to feel that regular naval personnel will replace reserves as soon as possible. But it will be a rich experience to be in on the ground work and to have seen Guam. . . .

Very best wishes to all,
Sincerely, Lillian

Agana, Guam
December 5, 1945

Dear Mother and Dad,

. . . We still don't know just when we will be home but think it will be by the last of January or so. At any rate it will be a much shorter stay than we expected. The school is progressing now so well that we can actually see it. But it has taken 6 months to get started. I hope they choose older girls who will really have an interest in it to take over. If anyone thinks we have not put blood, SWEAT, and tears into the project they are crazy with a capital C. Especially Miss Sears, the chief nurse. She is up many a morning at 3:30 or 4:00 a.m. working and planning or doing personal things that she can not get time to do during the day. Then she is up until 12 or 1 attending some (political) function that will get her what she needs for the school. . . .

Will finish this dull and uninteresting epistle as it is time for my corpsmen's class (9–10) p.m.

Love, Lillian

Agana, Guam
December 11, 1945

Dear Mother and Dad,

Just a quick note while I am waiting for the students to assemble. It is nearly 1 o'clock and I have had a busy day so far. At 8 supervision of the young nurses on the wards then finishing up the course outlines on Ethics and History of Nursing. Started to cut stencils for mimeographing that material. You see we have no text books for the girls so have to dig out the material and have copies made. Taught from 11 to 12, then lunch. Teach from 1 to 2 then back to the stencil cutting.

Cannot finish it today maybe tomorrow then begin work on Medicine, Surgery, and Psychiatry. That will keep me busy for two weeks or so, then maybe I shall have a few days with nothing but teaching and supervision load. I surely never thought that I should one day be a "supervisor" when I was struggling in my own student nurse days. Must close this as the class is here. We are studying the nervous system so yesterday I had the boys get a Guamanian toad and dissected it so at least they could see a nerve, the digestive system, and heart, lungs, etc. . . .

Love, Lillian

Beatrice Rivers served as a Navy nurse from 1943 to 1946. During ten months of that time in 1945 she was stationed on the U.S.S. *Relief* and was with the *Relief* at the invasion of Okinawa in April 1945. In September 1945, the *Relief* was ordered to sail to Manchuria to pick up recently released prisoners of war who had been incarcerated in Japanese internment camps. Beatrice Rivers describes that mission in the next three letters.

Dairen, Manchuria
[Aboard U.S.S. *Relief*]
Dear Mom and Daddy,                            September 9, 1945

. . . We had an escort of two destroyer escorts, which are slightly smaller than a destroyer and more maneuverable. They went slightly ahead and abreast of us and were mainly to sight and destroy all mines. We sailed at 6 p.m. and as soon as we were underway, our destination was announced over the speaker system as Dairen, Manchuria. Scuttlebutt then began anew. We were going to evacuate P.O.W.'s—all nationalities, women and children, return to Manila, no return to Okinawa, several trips, etc.

Our first day was uneventful. The second day out, excitement ran high. About 10 a.m. we went past a native boat which was almost under water and two figures in it waved excitedly. . . . They were Chinese, one young man and a young woman, suffering from exposure. . . . The natives are a mystery to us. No one speaks Chinese, so we fed them, and treated their ulcered legs. And now that we are in port, an interpreter came aboard and got their story. . . . They're originally from Soochow, near Shanghai. . . .

The next day was our most exciting. We passed through the mine fields in the morning and that was a cinch because we had the chart of where they were laid by our own bombers, but then we began to sight floating mines. The destroyer escorts would sight them, we'd all stop, and one of them would go alongside about 200 yds. away from it and explode it with their 20 MM guns. . . . It was so fascinating that one completely forgot danger. . . . All in all we knocked off seven mines. . . .

The next morning at 7 a.m. we sighted land . . . we edged into the big empty dock. As we got close, we saw armed Russian sailors patrolling. We docked safely and quickly, and an American officer came aboard and conferred with our skipper.

Here's the situation. The Russians came in and took over two weeks ago—mostly army. Our two destroyers came in a week ago. The Russians are in control. There's no type of American embassy or consular officials there. We may go ashore in small groups as "observation parties," officers only, for a brief interval in the a.m. or afternoon. We cannot go armed. No enlisted personnel allowed on shore. The reason is obvious—if there were any drinking and an "incident" developed, it would be very serious. . . .

The Chinese American Air Force in China has been dropping food supplies to the prison camp in Mukden and some of the ranking American prisoners flew down last night and came aboard. They told us that they'd return today and make all arrangements to fly down or send as many P.O.W.'s as we can take. . . . The P.O.W.'s are in good condition and will not be the critical malnourished patients we had anticipated. They will all be able to walk about. They are all thrilled to death and eager to get here. They are 200 miles away. We also eagerly await them. . . .

Love, Betty

Still in Dairen

Dearest Mom and Dad,     September 10, 1945

. . . The Russian authorities as I told you will allow a group of 4 to inspect the city as an observation party. So yesterday, Nita and I and four of our favorite doctors started to. We split up in groups of three and walked about half a block apart. . . . What a beautiful spectacle our lovely white ship presented as we turned and looked back. . . . One could see rats running along the wharf. We have shields on the cables that attach the ship to the pier to prevent the rats crawling up and on our ship.

We walked steadily for three hours. . . . In all a lovely city and typical of an average city of 700,000. But, every window was boarded up and very few people on the streets. . . . No stores open. The Russians are looting everything systematically and everything is stripped. Saw several fairly new automobiles stripped of tires and wheels etc.

and left by the side of the streets. What vehicles the Russians have are all USA vintage and they don't drive, they race them. It's strictly up to the pedestrian to get out of the way. Saw a few street peddlers sitting along the curb cooking hot sweet potatoes and some other unfamiliar stuff. Flies galore, etc. Very dirty. We finally got tired and decided to see the hospital and leave. . . . It's a 5,000 bed hospital and very modern. Built 20 years ago by Americans. Now it's still staffed with Japanese doctors and nurses. They were surprisingly polite, in fact, ingratiating. They took us thru the x-ray department, x-ray therapy dept. and operating room. Everything was filthy and the stench in the O.R. was worse than anyplace else. . . .

Love, Betty

Dearest Mom and Dad,                          At Sea, Near China Coast
                                                September 12, 1945

. . . Well, last night at about nine o'clock, our P.O.W.'s arrived by train. It took them 24 hours to travel 200 miles from Mukden, where they had been in prison. It was an experience I'll never forget. We had been anxiously watching all evening and suddenly the first handful came running down the dock waving their meager luggage. Our P.A. system broadcast martial music throughout the whole loading process. We had two corpsmen stand on the edge of the gangway, and they sprayed all passengers and luggage with DDT powder. We have all varieties aboard and mostly high rank. We have about 170 generals and colonels alone. In all about 800 passengers. We normally carry 621, so we have them on cots and stretchers, too. They are so grateful, they'd sleep on the deck. . . .

I kept dashing from my ward up to the main deck to watch, because my ward wasn't filled til last. . . . Finally my ward filled up. I have captains, 2cd and 1st lts., and nine enlisted men. 10 of my Lts. are B-29 pilots who were forced down around December, and in a separate camp. . . . Have seen Dutch and British officers and there are about a dozen Catholic French Canadian priests—mostly army officers and men and a handful of Navy and a few Marines.

Have heard atrocity stories which make your blood run cold. Apparently the worst hardship was while still in camp in the Philippines, and on board the dirty ships which brought them up to Manchuria. One group of 1290 men had only three hundred survivors after that ordeal. We have *Houston* survivors aboard—remember they thought only 8 survived. Well, they found 300 in a camp near Mukden. Even their families don't know this yet. What a wonderful surprise.

Our chief nurse is in 7th heaven. A dentist and doctor who worked with her at Bataan are aboard. She had thought one was dead and now thrilled to death to have them aboard. She's entertaining them now in one living room (ward room) with records and coffee. These men have so much to catch up with. They don't even know who Frank Sinatra is! . . .

I asked some if they got revenge on the Japs for cruelties and they said—no—We left it to the Russians. They ask no questions, and, of course, there were 2 or 3 quite kind Japs who did them a few favors, but they were very few. . . .

[Betty]


Occupation forces were a necessary part of the postwar world. Members of the WAC served in the Army of Occupation in both Germany and Japan. Emogene Heston of Ohiowa, Nebraska, received both B.A. and M.A. degrees from the University of Nebraska. In 1927, she married Vernon J. Moor, a veteran of World War I. He suffered a nervous breakdown in 1934 and was hospitalized for the remainder of his life. During the early years of the war, Emogene Moor worked in an aircraft plant in Omaha. In 1943, at the age of forty-one, she enlisted in the WAC and served in Colorado and Texas with the 3706th Army Air Forces as an Air WAC. In July 1945, Private Moor was sent to Europe as part of the Army of Occupation and was stationed in Berlin. While she was overseas, her husband died. In May 1946, she returned to the United States and was discharged a month later. After the war she owned a bookstore and also worked as a teacher in the University of Nebraska extension service. She wrote perceptive letters to her family and friends about the occupation of Germany during the first year of peace.[5]


                                                      [France]
My Dear Californians:                           July 18, 1945
    . . . Here I am in the huge forest of Compiegne near the town of the same name which was where the 1918 Armistice was signed. [Hitler also gleefully accepted the French surrender at Compiégne in 1942.] Headquarters in a beautiful chateau, beautifully landscaped, but with

5. The letters of Emogene Heston Moor are located in the Moor Family Papers, Nebraska State Historical Society, Lincoln, Nebraska.

very crude facilities. The chateau is empty but for G.I. stuff, and we live in tents and sleep on army cots and in an old bed roll and two blankets—more sometimes. . . . This is just a post from which they send people home, on to the Pacific, or to new assignments. It (for WAC) was just moved from Le Havre and with the PW labor, the camp will soon be ready for winter. . . . We had showers and shampoos in a G.I. wagon down by the river our first day in. . . . boy, after four days the shower felt good. We did our laundry in the river. . . . I am brown as a nut and feel so leathery. I hate it and wonder how it really was to live clean and snug but then these temporary things are gotten through, somehow, and we do have fun. . . . The G.I.s have been so glad to see us. We have had a good time, dances, tours and boat rides. . . .

The channel trip was smooth, . . . but the small boat was really crowded and lines for the latrines, which were made for men, remain clearly in my mind. Bunks 4 deep and all full, aisles narrow, for one at a time, only. We slept on board because ships still can't enter Le Havre in the dark. We saw 4 floating mines and they fired at 3 of them. Carl and I saw the first one real near and as the ship got past the guns let loose. Enough of real war impressions to last me. None exploded when hit, but I guess the Coast Guard would hunt them down. . . . Train service is still almost impossible in Germany. . . .

[Hochet, Germany]
July 21, 1945

Here we are in a suburb of Frankfurt—Hochet. The Army ordered people to move out in 24 hrs. and the army moved in—apartment buildings. . . . The first WAC in ransacked the place for things to furnish their rooms, but a few treasures were left for us. Four of us moved into a large room, third floor. It has a balcony and a large window which opens in. The china closet is a beautiful piece. We found plates, jars, glassware and vases to decorate the place. Pictures for the wall—potted plants for our boxes out on the balcony and window, a rug for the floor and we have a desk secretary and the table. Army cots with biscuits (blankets) (1 for sheet, one pillow and one cover). From attic to basement to garage we hunted and found our furnishings. It was like a treasure hunt. . . . Restrictions have been slapped on the whole German area over the week end. Rumor has it that there are isolated SS troops to be rounded up as well as some of our own AWOL boys. Billet to mess for us until Monday, so I'll get caught up. . . .

I love you all, Emogene

Berlin, Germany
My Dear Family:               August 22, 1945

This destination, I hope, will be my last for some time—I want to settle down and have time to assimilate what has happened the past three months. I arrived here the 18th, a month after coming to Germany. . . . The trip to Berlin lasted about two hours and most of that time we were above the clouds, . . . so I curled up on my baggage and finished my night's sleep. . . .

The army took over a section of Berlin in the American sector, little damaged. The offices are about 1 and a half miles from our billets, in the former Luftwaffe headquarters. The main building is quite impressive with marble entrance and elegant open stair way to the chief suites of rooms. Gen. Eisenhower will occupy one of these and his assist. the other in due time. . . .

The billets are blocks of three roomed apartments with six to a stairway. Two people live in each. The people were allowed to move out only personal things, and we have had to ferret out from basements and attics all the things we can find to fix up the places. The beds are hard old couches usually, and basements are full of packed dishes and treasures which we have fun unpacking and rummaging. We have found some nice things but lack a few essentials such as coffee pot, window curtains, radio, and wash board. . . . We feel like criminals digging out this stuff but keep right on doing it. We have found boxes of stuff which look as if it had been shipped in from somewhere else too. . . . The place is beginning to look very home like. . . .

Beginning Monday the boys do not have to carry firearms. They were so happy to get rid of them—it was such a familiar sight—everyone wore either a pistol or carried a rifle. . . .

Love Until another day, Emogene

 

Berlin, Germany
My Dear Family:               December 5, 1945

. . . Mills [another WAC] got into Potsdam last Sunday. She had an interesting experience so now I suppose I will not be satisfied until I succeed in getting there. The Russians keep all other nationals out but evidently the WAC are able to use their charms and get in. Mills delights to describe how the Russian GIs that brought them back to Berlin got some gas. They had 10 gal from a regular GI who had been out there on a date with a Russian WAC equivalent then when they got to town, they stopped at the motor pool. . . . Mills says she has never witnessed more clever flattery. She brought them to our apt. and fed them and they left very happy. She tried to take them into the club but couldn't. The place is really too small for our own detachment but it seems like a sad mistake to forbid our allies so emphatically.

I do hope they soon get every GI who fought over here out of the country. They just are not able to see the thing in the light suitable for peacetime understanding. The new Army boys may be able to make the adjustment. Fraternization is a horrible business at best. Gives me the jitters even worse than all the pro [prophylactic] station training and talks. Too easy to do wrong and then you'll be sorry lectures. Makes one ill, really. . . .

Love beaucoup, Emogene

Early in 1946, Emogene Moor was granted a twenty-day furlough, which she used to tour Germany, Switzerland, and Italy. She prepared a detailed travel log for her family. Her entry of February 4 described her attendance at the famous Nuremberg trial of twenty-two Nazi leaders.

[My Dear Family,]                                          [Nuremberg, Germany
                                                            February 4, 1946]

. . . Our minds were set upon the mission we had set for ourselves for this day—getting into the [Nuremberg] trials. At 9:00 promptly we were being checked at the Security Office and by 10:20 we were adjusting our earphones in the gallery at the trials. . . . We finally arrived to spend the first few minutes saying, well, there they are, just like the pictures. All 20 of the big shots. We stretched our necks to see each of them in turn. [Hermann] Goring with his head in his hand most of the time. [Rudolf] Hess really very alert this morning and paying close attention. [Alfred] Rosenberg not nearly as good looking as his pictures in the glorified days. (8 reels of Nazi films had just been shown in Berlin of the rise of the Nazis to power, Hess, Hitler and Rosenberg who had been the narrator of the story, were prominent figures in the parades, flag waving, band playing and great psychological effects and so much of the story had taken place in Nuremberg.) We did at last settle down to listen to the proceedings. The French were prosecuting the case for Holland and Norway. . . .

[Emogene]

Red Cross worker Jacqueline Haring was in the process of transfer from New Guinea to the Philippines when the war ended. She continued in her Red Cross duties in the Philippines until November 1945, when she was assigned to Kyoto, Japan. She was one of the first American Red Cross women to arrive in Japan after the war. She returned to the United States in the spring of 1946. (For other letters by Haring, see pages 194–196.)

[Somewhere in the Philippines]
Dear Mother and Dad,                                                August 6, 1945

The flight up was marvelous with a clear, blue sky and just enough clouds to make it fun. The plane was a regular parachuter plane with "bucket" seats and car and passenger all packed in together, but the kids I'd gotten to know at the strip made sure we had best seats and got us all tucked in our Mae Wests and safety belts before the take off and we left with a shout of good-bye which made me feel all kind of good inside. . . . We got to Leyte that evening and found out we were to spend the night at a casual camp but the pilots took us to their mess and that helped our morale a lot. After that they drove us to the town of T(?) [Tacloban] and I've never seen a filthier, more squalid, more mice-ridden town than that. It made me think of Phenix City! [a notorious town outside Fort Benning, Georgia, filled with beer joints and brothels.] The natives are of a low class and their houses are small thatched shacks with open fronts where they sell liquor, fruits and woven hats by candle light. There is no electricity in their part of town. . . . Of course, it was soon after payday. The soldiers were really doing the town—there, the natives "associate" with both white and colored Americans. Our abode was not attractive and perhaps it is just as well we didn't see it by daylight. It wasn't very clean and there was no light in the latrine or shower and all we had was my borrowed flashlight, but we were tired and it didn't bother us very long. There were a lot of civilians here who have been interned by the Japs and are still waiting to go home and they are quiet, white-haired shadows of people in the dark. We didn't talk to any of them but they slithered past me in the dark pathways. Then, there were USO girls with their tents filled with sequin-trimmed costumes which seemed to shake like diamonds in a coal pile. It was once described to me as the "black hole of Calcutta" and that wasn't far off. . . .

The trip up here [Luzon] was uneventful, but terribly interesting. The design of the terraced fields and the rice paddies was beautiful from the air and the bowl craters, wrecked planes, and shells of houses were strange to our eyes which had seen the war only beneath the camouflaging overgrowth of jungle.

The city [Manila] is all the horrible things anyone has ever said of it. Men fresh from Europe say the devastation is worse than any they have ever seen before. Every public building, private house, amusement building, apt. house, hotel is destroyed.

The houses along the path of the Japs retreat are marred with shell holes, machine gun bullets, scars and fire—there is nothing intact. The Japs buried and blew up all the beautiful classical and modern buildings before they left—there is nothing but rubble left. Marvels of cleaning up have been done and there is electricity here but there are still bombed automobiles on the side of the streets and activated bombs lying in craters of their own making in the sidewalks (one right in front of the Red Cross Area Hqs) and pontoon bridges across the rivers in the middle of the city for all the bridges were destroyed. . . . In the midst of all this the Filipinos live, plow their rice fields with water buffalo or walk the carabao down the streets stopping now and then to let them wallow in the rain-filled bomb craters. . . . Everyone wears wooden clog-like sandals. . . .

The small caratellas, or carriages, drawn by miniature horses are on the streets by the hundreds and mix strangely with tanks, ducks, trucks, jeeps, general's limousines, bicycle-pushed carts, man-pushed carts, and beautiful civilian cars which seem strangest of all. . . .

I'm delighted with [the offer of] my new job. . . . It's to be the largest Red Cross club in the world and will open about Sept. 15 in the building which was once the Jai lai Club and will be equipped with swimming pools, tennis courts, dance floors, restaurant, snack bars, library, craft shops, game rooms and anything you can imagine (even a tailor shop). We could get a cute place to live, have our own servants and work in luxury but I decided it just wasn't for me—that's not what I came overseas to do. So, I'm leaving tomorrow morning to another base about 175 miles north of here [near Lingayen Gulf] where there are troops still in combat, or freshly back for a short pass from combat where there's very little civilization, and where there might be work more up my alley. . . .

Much love, Me

[Somewhere in the Philippines]

Dear Dad,                                                    August 17, 1945

. . . It must be pretty gay at home now with everyone celebrating like mad and rationing ending and all. Over here, there's little reaction— no one seems to feel anything and as far as the hostilities being over are concerned, we are still a little dubious for we won't trust those Japs until they're safely occupied, disarmed, and imprisoned. Combat reports and rumors come in with news of peace and so we just say,

"The war is over," and wonder when if ever we'll feel like celebrating—inside of us. . . .

Much love, Me

[Somewhere in the Philippines]

Dear Mother and Dad, September 3, 1945

. . . It's finally beginning to seem as if the war is really over. Although I haven't had a chance to hear a radio here so that I don't even know what Truman had to say, it seems true at last. Watching the loads of men coming down from the hills now is perhaps the most convincing and the rising spirits among my men who had been slated to go in on the first invasion wave and remembered from other such experiences what that would mean to them have made us happy and relieved with them. . . .

Much love, Me

[Kyoto, Japan]

Dear Mother and Dad, November 3, 1945

Let me tell you about the trip. . . . We made our first landing at Lanay at the very northern tip of Luzon and had a leisurely cup of coke and doughnuts at the small strip canteen there. Then we took off and flew over heavy clouds to Okinawa. That was kind of fun 'cause we were flying by instruments, and it was quite a thrill to fly so many hours and minutes and then come down thru the clouds and find ourselves smack-bang over the one island (from the air) in that whole big part of the Pacific. . . . We were taken that night (after cooking a supper from 10 in 1 rations for the men and ourselves) at the plane, to a really transient camp for nurses and Red Cross girls. . . . We spent the whole day at the plane while they fixed the engines and we simply read, slept, cooked lunch and had a lazy day. It was restful but I was so anxious to get going that I nearly went crazy waiting. The next day, (yesterday) we finally took off. . . .

We flew over Ie Shima and Iwo Jima and a whole chain of volcanic islands (the Ryukus) until finally there was Japan! We landed at Kanoye and then we were on Japanese soil. . . . There were a few Japs around the field (which is surrounded by burnt-out hangars) and it seemed so strange not to see them under guard. Instead, they stood quietly by, staring at the American women and waiting to rescue any castaway food. . . . We took off again. . . . The weather was beautiful and the pilots made a special route to show us Hiroshima or what is left of it. We came down to 500 ft. or less (I could even see the spokes

in the bicycle wheels on the streets) and circled around and around
there, practically nothing standing, perhaps one or two buildings in
the whole city, and the entire area is a queer red color (which I understand is typical of the result of an atomic bombing). It seems impossible that people can live in the area, and still we could see people, cars,
and bicycles on the streets. I feel as if I should somehow explain it so
you can see it but there just isn't anything to say except that there
isn't a thing there. . . .

After that we climbed over the mountains and flew over Kyoto and
Biwa Lake and I waved to my 33rd Division as we raced along. That
city hasn't been bombed at all and looked like a park from the air—
all green trees and carefully arranged buildings . . . finally there was
Fujiyama piercing into the sky. . . . After we'd passed it, we could
look back and see it against the sunset sky and it was beautiful. . . .

I'm on an overnight train to Kyoto! And here's the most amazing
thing of all, I'll be the *first,* in fact, the *only* for awhile, girl there and
the gem of all assignments here in Japan (or the Southwest Pacific, for
that matter) and I got it! Me! . . . The man area director has just
half-way opened a club and canteen in Kyoto and that will probably
be mine. . . . I imagine I will live in a hotel until there are other girls
so we can live in a billet. I wonder if I'll have an MP outside my door.
It'll feel like St. James Palace.

Much love, Me


                                                         [Kyoto,] Japan
Dear Mother and Dad,                            December 16, 1945
I started a letter several days ago but have been so terribly busy I
haven't ever finished it yet, so I'm dashing off this note. . . . I think
tomorrow we will finish the club and start Xmas work. We opened
the theater with a USO show Friday night and it was a huge success
(even the Col. said so). The stage looks wonderful with its red velvet
curtain and red and silver backdrop, and all the footlights and flood-
lights. I'm proud of it even if I haven't figured out how to draw the
curtains from one side yet.

Hope I can work out a good Xmas program. I've got to wrap
3000 Xmas presents, too! We're having four trees and making all the
decorations and the Santa suit. There's a lot to do.

Must go to sleep now for tomorrow will be another hard day. Will
write again soon.

Much love, Me

[Kyoto, Japan]

Dear Mother and Dad, January 26, 1946

I just wrote what should be my last letter of these many months and I hope this will be my last word to you besides telegrams and telephone calls until I reach Detroit. You see we have heard that we board the ship tomorrow and I *think* it's probably true. . . .

We'll come post-haste home—but, please no big parties, in fact nothing planned—just everything as it was with marble cake and lots of fresh milk in the ice box. This is my last letter of this series.

Much love, Me

WAC Lieutenant Martha Alice Wayman, a civilian personnel officer, was also stationed in the Philippines and Japan during 1945 and 1946. In fact, she was one of the first WACs sent to Japan after the war. Her letters to her mother provide additional information about the occupation of Japan in the months immediately following the war. (For other letters by Wayman, see pages 181–182.)

Manila

Dear Mother, August 1, 1945

. . . Today we moved *again*—this time to stay awhile, I hope. Our office is rather messy but should get straightened out sometime. It's located in a burned building in what used to be the fashionable shopping district of Manila—the Escolta. The building itself isn't in too bad a shape now that they have all the holes patched. But right beside it, leaning in the opposite direction, however, is a building which is turned almost on its side. They will probably tear it down sometime before it falls on the other building. . . .

What's new back in the States? Are things as hard to buy? I guess I'm better off over here now—no worry about rent, sergeants, food, gas and light bill (we don't have either gas or lights though) and no ration coupons. So I guess I'll just stay awhile. . . .

All for now, Martha Alice

Manila

Dear Mother, August 11, 1945

. . . Our office has been in a turmoil all day no one knowing just what to do or what to believe. We heard the news at about 10 o'clock last night but no one knew just what the score was. Then today we saw all the headlines in the papers and we still didn't know what to believe. . . .

[Martha Alice]

Dear Mother,

Manila

August 13, 1945

We are all still waiting to see what is going to happen. We've already had our celebration, so we won't worry about that. We celebrated the second rumor but not the first. . . . Anyway, around midnight we saw the searchlights in the sky and the rockets and everything. We didn't pay much attention to it until Cecil and Marjorie called attention to it. . . .

Martha Alice

Tokyo, Japan

[Dear Mother],

October 21, 1945

. . . Here I am, in Tokyo—after almost 18 months overseas. This is quite a contrast to Australia, New Guinea, or the P.I., but a very pleasant one. . . .

We flew over Tokyo and could see the destruction caused by Allied bombing. It didn't look much worse than Manila but it was a different type of destruction. Whole areas would be destroyed but with a few large buildings left standing amid all the ruins. Since most of it was done with incendiaries, the fireproof buildings escaped quite a bit of the damage. . . . The streets are less torn up than Manila and street cars are still running. . . .

We took the longer way round (from the airport) going through Yokohama. The roads were not paved until we got almost to the city itself. All along, we could see civilians walking, riding bicycles, standing in the fields and yards, many of them stopping to stare at us as we went by. A lot of them would wave at us or smile, and a few of the little kids would make, of all things, the V for victory sign! . . .

I got a pretty good view of the destruction in the city. Many places were completely destroyed—for blocks at a time—and others were scarcely touched. . . . We drove past the palace with its high walls and moat. . . .

At the present time there are just 7 WAC officers in Japan but 60 more are coming from the states very shortly. They will be in the censorship, also, having been trained for several months in the States. . . . I'm with the civilian personnel again, as Civilian Personnel officer, with a second Lieutenant as Labor Procurement Officer. We will both be doing about the same kind of work but mine will be for . . . Japan and his will be for the Tokyo area only. The work should be very interesting. . . .

Martha Alice

                                                              Tokyo, Japan
Dear Mother,                                                  May 9, 1946

Right now I'm down at the office, supposedly working, but it happens that there isn't much to do. Officially I'm relieved of duty here but have a few things to get straightened out before I go out to the Re-placement Depot (Repple Depple). . . .

I'll let you know definitely when I'm leaving and on what ship. . . . It looks like I'll be stopping in Seattle instead of San Fran-cisco. . . .

I think this will do for the time being. This machine isn't too easy to write on, so I think I'll stop. I guess I might as well go home—nothing more to do around the office at this late an hour. All for now.

Martha Alice

With the end of the war, the return of millions of Americans from their overseas stations to the United States proved to be an immense task. Vessels of every description as well as many types of aircraft were assigned to the duty of bringing home the troops. In the next selection of letters, WAVE June Kintzel, a flight orderly, reports to her parents about her part in the Stateside portion of these journeys. As a flight orderly based at the Naval Air Station in Olathe, Kansas, she often accompanied sick and wounded troops, recently evacuated from over-seas, on cross-country airplane flights to hospitals in the United States.

                                                           [Columbus, Ohio]
Dear Folks:                                                   May 16, 1945

. . . [Norfolk] is where the 11 patients got off. They'd just come back from Iwo Jima and had spent a little time at the Naval Hospital in San Diego before being transferred to the hospital at Norfolk. There were 7 Marines and 4 sailors all of whom could navigate by themselves. Two of the men were on crutches with gunshot wounds in one leg. The others had fractured or bullet-wounded arms. They weren't any different from regular passengers. They did make it clear that they thought V-E day was a lot of baloney, because they said the war was far from over. . . .

On hospital flights we have planes that are equipped to carry stretchers if necessary. They'll hold 24 of them and they're piled four deep. We had eight of them up for the fellows to sleep in if they wanted to. The rest of us sat on "bucket" seats, the kind that fold out from the side and the paratroopers use. They're called buckets because they're scooped out to allow for parachutes to be sat on. . . .

Lots of love, June

[Olathe, Kansas]
Dear Folks:                                        August 8, 1945

. . . Going west I had regular passengers, but coming back there were patients from San Diego on. Of course 2 stayed at Phoenix that night and then had a different batch of patients from there on in. Eleven litter and four ambulatory cases made up the load and some of the litters were really bad off. Four of them looked like skeletons—being nothing but skins and bones, and having awful colors. One of them was dying of blood poisoning. Some of the others had both legs in casts, while another Marine was paralyzed from the chest down. A sailor had had both arms and face burned, but he didn't look too bad. There was a doctor, a nurse, and a Corpsman along, so the only thing I did in the way of taking care of the patients was to get them a drink of water, or talk to those who felt like talking. Most of them were back from Okinawa. I had heard of the way the planes got to smelling when there were such cases aboard, but it wasn't too bad as I soon got used to it. Now you know why I'm not overjoyed about having another one, even if you do get more time off in Oakland. I was one of the first girls to take a hospital flight west as the fellows have been doing it. . . .

Lots of love, June

Olathe, Kansas
Dear Folks,                                        August 16, 1945

It hardly seems possible that the war is finally over. It seems like I practically grew up with a war going on, and therefore it should keep right on going. . . .

I wondered what you folks were doing Tuesday night in the way of a celebration. . . . I had arrived at about 9 AM from my trip west, having been up then about 24 hours. I took a little nap (all of 3 hours) and then we decided to go into Olathe for dinner. As we stepped off the bus, everyone was yelling that the war was over. We didn't get very excited as it was the tenth such report, so we went in and ate. In the meantime every whistle in town (all of two probably) started blowing along with honking of car horns. Where all the people came from in a jiffy, I'll never know, but they started parading around the courthouse square, preceded by the high school band. That's when we began to get excited. . . .

Perhaps you've seen or heard about the Navy's point system. For Waves a total of 29 points is needed for discharge soon, or before six more months. I have exactly 15. We orderlies figure we have about 6 or 7 more months, but you can never tell. Most of the kids here aren't too eager to get out, mostly because they won't know what to do, and they don't mind Navy life. I feel the same way, although it will be swell to go home. . . .

This last flight west we returned from Phoenix via El Paso and Ft. Worth, but I didn't get to see a thing as that part was made in the dead of night. As I mentioned before it was another hospital flight, but all of the patients were ambulatory. . . . I [don't] like hospital runs because of the general condition of the fellows, the odor, the great amount of extra gear I take, and the uncomfortable planes we use. They really don't have a very good system worked out for patients. I know I'm tired when I get back, and I only travel half as far at once as they do and I'm healthy. I think they must be exhausted by the time they reach the east coast, besides being hot and dusty and in need of a shave. . . .

This makes four full pages and I'm all worn out. . . .

Lots of love, June

By the early months of 1946, June Kintzel began to give serious consideration to her life after the WAVES. In the next letter, she tells her parents about her plans to go to college. Kintzel, like thousands of other servicewomen, made good use of the G.I. Bill to further her education.

Olathe, Kansas

Dear Folks                                    January 19, 1946

. . . Another decision I made this past week was what I'm going to major in college. It's meteorology. I figured with some scientific subject I'd stand a much better chance of getting work than with a general course in airline transportation. With meteorology I could go to work either for the airlines or perhaps the government. There are a lot of fellows working here who formerly were with the airlines, and that's what they advised, too. One fellow also said, when I asked him if I'd be a little late in the game 4½ years from now, that he thought they'd just be going full-swing by then. Monday I'm going to see the education officer and find out which universities have good courses in it. I understand the U. of Chicago has one of the best, but I want to find out which other schools do, too. I also think I'll go see the WAVE officers who are meteorologists with Naval Air Transport Service here. I wish I were at home to talk it over with all and sundry, but I guess I'll have to make up my own mind. "I'm a big girl now, eh what?" . . .

Lots of love, June


June Kintzel was discharged from the WAVES in July 1946. In September 1946, she entered the University of Wisconsin, where, three years later, she received the B.S. degree in geography. She then obtained a position with the Central Intelligence Agency as a cartographic compiler in the Geography Branch, where she met her future husband, Dean Rugg. June Kintzel and Dean Rugg were married in December 1951. They live in Lincoln, Nebraska, and enjoy visiting with their three grown children, who live in widely scattered parts of the United States.

MARIANA "NANCE" OLWEILER of Elizabethtown, Pennsylvania, joined the WAVES in April 1943 and graduated from Yeoman School in Cedar Rapids, Iowa. She was stationed in Corpus Christi, Texas, where she worked in the personnel office until her discharge in April 1946. She, like June Kintzel, planned to use the G.I. Bill to go to college. In the following letters, she discusses these plans with her parents.


Dear Folks—
Corpus Christi, Texas<br>November 15, 1945

. . . I have done nothing definite about school such as enrolling anywhere. Dad, if Mr. Klouse is still back at school, I wonder if you would be able to contact him and find out if he has any suggestions as to a place for me to go. I can hardly make any definite moves because I don't know enough about the courses in music that different colleges have to offer, and it is too big a task to begin writing to so many of them. I think he would possibly have a few good suggestions to make. Anyway, it is worth a try. What you think? . . .

Love, Nance

Corpus Christi, Texas

Dear Dad,                                              February 6, 1946

I was up talking to our education officer yesterday, and she advised me to get two copies made of my high school transcript, one of which will be sent to the school that accepts me, and the other which will go to the Veteran's Administration in Washington. I am certain now that I will not sign over until 1 September, for she told me that if that were the case, I would not get started to school until the following February; the reason being that after my discharge, it may take as long as three months to send in my transcript, copy of discharge, letter of acceptance from the college, etc. to the Veteran's Administration and receive answer from them. . . .

I got your letter last week telling of the $500 that you salted away for me, and of course, I can never thank you enough in mere words. . . . Gee, Dad, when I think how you went thru college—and I've thot of it many times—I honestly and sincerely appreciate the things you have done for me, and how that it will be so much easier for me as far as money is concerned. You had no Dad to help build up your bank account in a big way such as you are doing for me, and you had no G.I. Bill of Rights to see you thru. . . .

Lovingly, Nance

After she received her discharge from the WAVES in April 1946, Mariana Olweiler used the G.I. Bill to study music at Peabody College in Nashville, Tennessee, and at the University of Washington in Seattle. Late in 1947, she married Harold Allen, a Marine war veteran, who was also a student at the University of Washington studying with the support of the G.I. Bill. The Allens have three children and one grandchild. Mariana and Harold Allen live in Elizabethtown, Pennsylvania. Mariana Olweiler Allen has maintained a lifelong interest in music and has performed with several community orchestras.

WITH THE WAR'S END, many uniformed women looked forward to getting married and raising a family. Anne E. Ferguson, a Red Cross clubmobiler who went ashore on the Normandy Beachhead on July 18, 1944, and subsequently followed the troops through France, Belgium, and Germany, provided a wealth of details about her forthcoming marriage to Lieutenant Colonel Pierre Boy in letters written to her parents during the spring and summer of 1945. As these letters demonstrate, the traditional rituals of life continued—even for women who had been very close to the front lines of battle. Anne and Pierre Boy now live in Nottingham, New Hampshire. They have five children and seven grandchildren.

                                          [Muhlhausen,] Germany
Dearest Mother and Daddy,                       May 13, 1945

Back "Home" after our trip to England—but this letter isn't going to be about that. I suppose that during the course of these past months you may have noticed my mentioning a certain person by the name of Pierre Boy. Please cease to consider him as just a person, because he is going to be your son-in-law. . . .

He was one of the first people I met when we went to Verdun last September. I liked him immediately, and as things will, one thing led to another. Ever since February, we have really known how it was going to end, but I guess we were trying to be sensible and put off the ultimate! This morning we signed the necessary papers and turned them in for General Doran to sign, too, because I will definitely have to stay over here for at least 6 to 8 months and because Pierre is going to stay in the army, we decided not to wait any longer but to take what happiness we could get as soon as possible. The waiting period is two months for the army, and I have to become a volunteer for the ARC, with the exception that my maintenance will be paid.

I am trying to think up all the answers to all the questions you will want to ask. First of all, I love him and to put it mildly—it's mutual.

He was born in Canada on April 15, 1914. His mother is French and his father Canadian. . . . He now lives in Berlin, N.H.—went to p.s. there and graduated from the University of N.H. in 1939 (degree in forestry). On graduation he went right into the army as a 2cd Lt.

and has had 50 months overseas in that time—Trinidad and Panama and England, and here. Present rank, Lt. Col. . . . He is 6 feet tall, brown eyes and his hair is graying—was almost black. . . .

He has had to work hard all his life for his people are far from well off. This accounts for the date of his college graduation. He worked for a couple of years first. . . .

The only dark note to the whole thing is that you won't be here for me to talk to and worst of all won't be here for our wedding. I have never been able to picture myself in traditional white, but I had always supposed that you, Pops, would "give me away" and you, Ma, would be there to weep or something. I hope that you feel that we should go ahead. The future will always be indefinite, and as long as we both have to stay here, we figure we want to do it together. . . .

Heaps and heaps of love, Anne

Muhlhausen, Germany

Dearest Family:      May 27, 1945

. . . We have definitely set the date for the wedding for July 17. Before the chaplain can do the job, we have to go to Luxembourg to have a civil ceremony, because Germany has no civil government. How I am going to get all the things done that have to be done, I don't know. Just for instance, my summer uniform has to be thoroughly altered before it is wearable, and it is in Weisbaden and I'm here. Jane and Mary have decided to take a firm hand and even plan a reception. Jane said she never had a chance at her own wedding. Her mother did it all, so now she is going to town!! There is no such thing as a quiet wedding when you have friends in the vicinity, I guess. . . .

Heaps and Heaps of Love, Anne

Muhlhausen, Germany

Dear Mother:      June 22, 1945

The enclosed list is only of the people I thought you might forget or not have addresses for. You see, I am assuming that you plan to send out some form of an announcement of the wedding. I figured that you would think of all the people in Schenectady, both young and old, who ought to be included. . . .

Heaps of Love, Anne

Eschwege, Germany
Dearest Family:                                              July 6, 1945

This past week, we have been very busy moving and getting set up in the new location. The Russians pulled into Muhlhausen as we moved out. The Germans had been streaming out of town for days to get out of the area. They expect the same sort of treatment they dished out at Nordhausen and Buchenwald, I guess. . . .

The most exciting thing I have to tell you is that the package with the nightie, negligee, slips and bathing suit arrived last week. The things are beautiful, fit perfectly, and are very becoming. The red dress will be wonderful at the Riviera, as it is really hot there now.

I had been expecting to pad around in my bare feet or else in my tan (and dirty) slipper-socks, but Mrs. Weir (Jane's mother) came through with a swanky pair of blue satin mules! Jane also had her mother send me a lovely white nightie, so with the pretty yellow one and a blue one Aunt Anne sent me for Christmas, I am pretty well fixed. . . .

Heaps of Love, Anne


Wiesbaden, Germany
Dearest Mother and Daddy:                                   July 16, 1945

. . . This afternoon we have a rehearsal of the wedding, just as if we were having one at home. Bob Abbott, who was to be Pierre's best man, was sent to the USA today, so George Shine has been promoted from usher to fill in, and we have added Judge Maynard to the usher's ranks. Tonight the boys are taking PDB in hand and are, I know, going to try to get him drunk. He will be utterly miserable, as he is a most moderate drinker, having been tight probably twice in his life, if that. Louise Smart Emmons is having me and all the girls up for a hen party after supper. I hope it will be brief so that I can get in a good night's sleep. . . .

Heaps and heaps of love to everyone, Anne


Anne Ferguson and Lieutenant Colonel Pierre Boy were married in Weisbaden, Germany, on July 17, 1945. Anne Ferguson Boy wrote the next letter to her parents while honeymooning in the Riviera.

Cannes, France
Dearest Family, July 20, 1945

As far as I can see, the wedding went off without a hitch, Jane got rather excited when the organist hadn't been located by two o'clock. But he had by that time been practicing in the church for over an hour. Lots and lots of pictures were taken, and they ought to be ready when we get back to Wiesbaden. We had a nice reception—General [Omar] Bradley came and our own 5th Armored General Oliver plus Brig. Gens. Doran, Kibler and Moses. Pierre and I felt very honored, though it was a tribute to the ARC work over here more than to us personally.

. . . We spent Tuesday night in Wiesbaden in a luxurious room in one of the hotels, then Wednesday a.m. caught the plane down here. It was a gorgeous flight—right over the Swiss Alps. There were glaciers and patches of snow and pools of water either black as ink or blue as turquoise. Lake Aggasiz was most beautiful—green fields along the shores and the snowcapped Alps rising sharply from the blue-green water. . . .

Yesterday we strolled around Cannes and in the evening went to a movie and then a little dancing. . . . We have a suite (!) in the Carlton Hotel. The army certainly treats its honeymooners well. The rooms are on the front of the building, so we get the sea breeze and the sound of the waves on the shore. This life is wonderful! . . .

Heaps of Love, Anne

Preparing for the postwar world also included raising a new generation of Americans. The following January 1946 serial letter, written by Miriam "Muffett" Dawson Hudson of Providence, Rhode Island, to her Army husband as he traveled back from his tour of duty in Japan, provides an intimate glimpse into the events leading up to the birth of a "postwar baby boomer." Miriam Dawson enlisted in the WAC on June 7, 1943, the day after her twenty-first birthday. She was assigned as a telephone operator at Marana Air Force Base near Tucson, Arizona, and became acquainted with Bill Hudson, an enlisted man in the Army, while making a long-distance call for him to his parents in Grenola, Kansas. They fell in love and were married on February 26, 1945, at Camp Howze, Texas, where Bill Hudson was in training. She then returned to Tucson, and he was sent to Europe. After the European war ended, he returned to the United States on a thirty-day delay-en-route furlough on his way to the war in the Pacific. During this furlough, Miriam Hudson became pregnant. She was discharged from the WAC in September 1945, and she returned to her childhood home in Providence to await the baby's arrival and her husband's discharge from the service. Bill was discharged in February 1946, and their son was born on May 2, 1946. Bill died in 1970. Miriam Dawson Hudson lives in Winfield, Kansas. She became Commandant of the Winfield, Kansas, Post of the American Legion in 1988, the first woman commandant of this post.


                                                [Providence, Rhode Island]
My Darling:                                      January 16, 1946

I'm going to start a day-to-day letter to you, which I will send when I get word you're in the States. . . .

Do call me, as soon as you can, when you get to Grenola! Just hearing your voice will do so much for me; I've missed you so! I never told you how much, because I didn't want to worry you, but the need for you, at times, has almost been a physical ache. I can hardly wait for you to take me in your arms, and we can tell each other all the things we've written,, so many times, I love you, Darling! Your own, forever, xxx "Muf" xxx


Bill, Honey!                                     January 18, 1946

. . . "Our baby," has been on his best behavior, since he heard his "Daddy" was coming home! Guess, maybe, you've got him trained, already! No kidding, darling, the nearer the day comes, the more excited I get! "It" is so lively, kicking and moving around all the time, I can hardly believe that this is happening to me. . . .

Your own, "Muffet"


                                                 January 21, 1946

. . . Well, honey, your pregnant wife is getting weary, sitting up, so I think I'll lie down, and rest, awhile! Good nite, darling, sweet dreams, I love you!

Your own, forever, "Muff" xxxxx


                                                 January 23, 1946

My Honey! . . . Our baby is so lively, darling, wait till you see how "he" jumps around! Oh, yes, you can see "him" move. In fact, he's almost kicked things out of my hands! . . .

January 31, 1946

. . . Well, am hoping to be really Mrs. Wm. Hudson, again, for keeps, around the 15th of February! And, honey, it will be marvelous! Want to see you as soon as possible, *but* I'm not going to be selfish, I want you to spend some time with Mom and Dad, before coming East, 'cause they won't see us till August, again, and they've missed you as much as I have! I know you'll be as anxious, as I am to have you to get here, but we've got to share our happiness, with others! . . .

Damn the Army, anyhow! By the time you get out, and home, and then, to Prov., I won't be able to *love* you the way I want to, 'cause you can't have relations with your husband for *two* months *before,* and *two* months, *after! Ain't* that awful? Gosh, I get all "gooey" just thinking about you and how nice "it" would be, what will I do, when you get here, and "it" will be so near, and yet, so far!! (Fresh, aren't I?) Guess it's a good thing, we'll have twin beds, after all, 'cause I couldn't stand it, otherwise. . . .

Well, my honey, I want this letter to beat you home, so I think I'll close, and *patiently* (am I kidding?) wait to hear your voice, soon, I hope! I love you, very much!

Your own, forever, "Muf" and "Dimples"

XXX       XXX


Not everyone could face the future as blithely and happily as Miriam and Bill Hudson did. Some women veterans would be sorely disappointed as they searched for employment commensurate with their wartime skills and training. All too often, they could only find low-paying, unskilled, and semi-skilled jobs. African Americans had hoped that victory over fascism abroad would be matched by victory over "Jim Crow" at home. Although President Harry Truman issued Executive Order 9981 ending desegregation in the armed services in 1948, changes in civilian life occurred much more slowly and grudgingly than in the military.

For lesbians in the military, the crusade for equitable and impartial treatment was just beginning. Writing to *YANK* in the fall of 1945, a former officer in the WAC offered a forthright and prescient appeal for tolerance and understanding of homosexuality.[6]

---

6. "Mail Call," *YANK,* November 16, 1945, p. 18.

Columbus, Ohio
Dear *YANK:*                                           [Fall 1945]

I firmly believe recipients of discharges other than honorable are, in their own unique way, branded for life. I feel I know wherein I speak, for my case is one of many who received the in-between discharge "Without Honor". . . .

To be more explicit, I served for approximately three years as an officer in the WAAC and WAC. My efficiency ratings were "Excellent," and in due course, I won my promotion. Then I broke—in a moment of insanity wherein another WAC suddenly attracted me in my lonesomeness. This placed me in such a mental and spiritual upheaval that I requested Washington accept my resignation from the service. After four months of debating, it was accepted. . . .

At the present moment I am trying to land a decent job. Since I entered the WAAC immediately after graduating from college in '43, I need must show my discharge in order to account for the three intervening years; also most application forms require information on military service. The result is—I am still hunting for a job.

The public in general is uneducated in the psychology of handling my type of discharge, hence I find it embarrassing and impossible to elucidate upon just why I left the WAC—and just what does this type of discharge mean? Many army medical doctors believe strongly concerning the injustice of this situation. If only people would realize this and help us with understanding rather than casting us out with condemnation!

I use the word us, for I have voluntarily drunk from the Lesbian cup and have tasted much of the bitterness contained therein as far as the attitude of society is concerned. I believe there is much that can and should be done in the near future to aid in the solution of this problem, thus enabling these people to take their rightful places as fellow human beings, your sister and brother in the brotherhood of mankind.

[Name Withheld]


As the letters in this book have demonstrated, United States women in uniform frequently wrote about the meaning and significance of their wartime experiences. What follows are three sets of letters that offer personal and powerful commentary about the experience of war and its larger meaning.

Gysella Simon, a Red Cross club director in England and Europe, began to evaluate the significance of her wartime experiences more than a year before the war ended. Her letters are addressed to a friend and to her parents. (For other letters by Simon, see pages 155–157.)

[Somewhere in South Wales
United Kingdom]

Dear Mil:                                  May 21, 1944

Your letter of May 8 touched me deeply, for I know how lonesome you must be. I just wish your life could be as full of romance and adventure as mine is at present. I often think of you and wish you were here with me. When I think back on the pathetic creature I was about a year ago, I can scarcely believe it At last, here in this forgotten place, I have found myself. For 5 months I have lived with men preparing for combat. I have eaten with them, laughed with them, cried with them, have shared their fears and anxieties, hardships, etc.—and wonder why I didn't get in this sort of work sooner. Of course, I have my moments, too, but most of the time I feel I am doing a real worthwhile thing and it makes me glow with satisfaction. I should like to share this feeling with every American girl back home, but most especially you, for I know how you feel. I know too, that without you at home, I should not be here, so in a way you are making a real contribution and you can be sure I shall never forget it.

Life is very simple now. Our men have gone forth and we are all alone, Virginia, Trish and I, all alone on this windswept hill that only a short time ago was teeming with last minute preparations for the Invasion. During the last few days before they departed I spent a good deal of time with the Rangers, those daring men who will be in the 1st wave, no doubt. What a brave lot, they are, Mil—so young, so strong, so handsome and eager for battle. I thought of all their mothers, sweethearts, wives,—how proud they must be. How can you expect me to be the same Gizzie that left 5 mos. ago? I know deep down I have changed a lot already and wonder what the rest of the duration will bring? . . .

Love, Gys.

[Somewhere in England]

Dear Folks:                               August 8, 1944

With the war in Normandy going so well we are at least able to see the "beginning of the end." You'll find me much changed when I get back, not physically except that I'll be much slimmer but rather mentally. To have seen what my eyes have seen; destruction and devastation far above and beyond the scope of your comprehension; casualties and survivors; hunger and privation, naturally have left a deep impression on my mind. I know I shall never be able to forget it . . . we shall return a strange lot who will need every ounce of understanding you are capable of giving us. . . .

On the other hand, we'll be so damned glad to get back, we may forget easily—but I doubt it. Time will tell. Of this I am sure—as long as I live there will be no room in my heart for people who are small in thought or deed for I have had the great experience and honor of working with *heroes* who come from every walk of life. They are truly the great people of the world today. . . .

May God Bless You All, Love, Gys.

Dear Folks—

Belgium<br>January 1, 1945

. . . I thank God for keeping you from all this, but secretly and facetiously, I feel sorry for you for missing it, too, for if I get back I'll be the richest, most grateful gal in the whole U.S.—grateful for such things as water running from a faucet, Radiators—heat—bright lights, soft, clean beds—Quiet sky. I'll be rich in good fellowship and kindness. God, I'm lucky. And how damn proud I am of my outfit. You couldn't drag me away with a $50,000 bribe. We're all together a small army of Americans sticking it out as all of you back home would do if you were here. Sticking it out here in the hell of Belgium today and able to smile and laugh (in between alerts) and crack jokes. You can't beat American humor—no matter what. . . .

Gizzie

In the autumn of 1945, Faye Anderson, a Red Cross worker in the Philippines, wrote a letter to her co-workers in New Guinea in which she summed up her feelings about the war. (For other letters by Anderson, see pages 197–198.)

Dearest Ones at 1140:

Leyte, Philippine Islands<br>October 28, 1945

. . . What is the matter with people in the States, having strikes, strikes, strikes. . . . Don't they realize we all have to pitch in and try to save this Peace which has cost so dearly in human lives? I'd like to take some of those strikers and transplant them over here in the jungles and mud for a while and maybe they'd sing a different tune. . . . If they had to live for months on end in mud, filth, eating C rations, see their best friends die, and be sick with fevers, jungle sores and all the rest, maybe they would see things in a different light and give our doughboys a chance when they get home rather then sending them back into a world of petty strikes and conflict. . . . The way we are carrying on now, another war will be on our hands in 25 years. Excuse this rambling, but it all burns me up. I do not pretend to know a thing about business and politics but I've seen what this war has done over here and I know if I were a returning GI I'd be mighty fed up with Our Country I'd been fighting for. . . . I'll never cease to be amazed at the things we have done over here—the heartaches as well as the good experiences—but I guess you know by now that I love this sort of work and it has been a priceless experience. . . .

Loves 'ya, [Faye]

In the final selection, Dorothy "Dode" McDonald, a Red Cross worker who spent thirty-two months in Europe, draws perceptive conclusions about her years at war.

*USAT Christabal,* At Sea
August 30, 1945

Dearest Mother and Dad—

This is the last of my overseas letters. I suppose I should recapitulate neatly, draw a few succinct deductions, and close with a flourish, but I'm not quite up to it, I'm afraid. I feel somewhat like the weary sky fighter, homeward bound from a last flight, consumed with a desire to get back safely, write "mission accomplished" on my briefing report, and throw myself with frantic relief onto my bed. Oh! to sleep and sleep and rest and rest! Tho' I never wish to forget. Never for one moment the men and days which have made these two years and a half forever mind and heart stirring. History can recapitulate. I will remember. . . .

It has been a tedious voyage. But now the interminable hours are counted and done and we are lying quietly at anchor in Boston Bay. It has been dark for hours following a sunset which died a scarlet death over a nearly motionless sea. We've stood by the rails in the dark and peered into the grayness ahead searching for the single light that would suddenly mean America and Home! It's been so long. And then we saw one light in our world of gray and the soldiers yelled and

shouted and whistled and cheered. And another light appeared. Then a silence came over us all as a ring of lights leaped twinkling into view. And each of us took his own thoughts and held them quietly in heart for endless moments, watching the lights, and coming home. . . .
Dode

# For Further Reading

No book-length study of the history of women in uniform during World War II has been published. The best introduction to this topic may be found in D'Ann Campbell, *Women at War with America: Private Lives in a Patriotic Era* (Cambridge, Mass.: Harvard University Press, 1984), chapters 1 and 2. Campbell has published several excellent articles on servicewomen and World War II. Of special interest is "Women in Combat: The World War II Experience in the United States, Great Britain, Germany, and the Soviet Union," *The Journal of Military History* 57 (April 1993): 301–23. Another good general study of uniformed women during the Second World War is included in Susan M. Hartmann, *The Home Front and Beyond: American Women in the 1940s* (Boston: Twayne Publishers, 1982), chapter 3. Information on African Americans in the armed forces during the war years appears throughout Jesse J. Johnson, *Black Women in the Armed Forces, 1942–1974, A Pictorial History* (Hampton, Va.: Hampton Institute, 1974). An older work based on the letters of women on the war fronts is Alma Lutz, ed., *With love, Jane; Letters from American Women on the War Fronts* (New York: John Day Company, 1945). Articles about servicewomen and World War II may also be found in *Minerva: Quarterly Report on Women and the Military*.

The standard work on the history of the Women's Army Corps is Mattie E. Treadwell's pioneering study, *United States Army in World War II, Special Studies, The Women's Army Corps* (Washington, D.C.: Department of the Army, 1954). This may be supplemented with Anne Bosanko Green, *One Woman's War: Letters Home from the Women's Army Corps, 1944–1946* (St. Paul: Minnesota Historical Society Press, 1989). Also valuable is Leisa Meyer, "Creating G.I. Jane: The Women's Army Corps During WW II," Ph.D. dissertation, University of Wisconsin, 1993. A number of memoirs by World War II WAC veterans have recently been published. For the experience of a WAC stationed in the United States, see Catherine Bell Chrisman, *My War, W.W. II—As Experienced by One Woman Soldier* (Denver, Co.: Maverick Publications, 1989); on the European experience, see Dorothy R. Spratley, *Women Go to War: Answering the First Call in World War II* (Columbus, Oh.: Hazelnut Publishing, 1992); on the Pacific experience, see Nancy Dammann, *A WAC's Story: From Brisbane to Manila* (Sun City, Ariz.: Social Change Press, 1992). An older memoir is Auxiliary Elizabeth R. Pollack, *Yes Ma'am: The Personal Papers of a WAAC Private* (Philadelphia: J. B. Lippincott, 1943). A novel with significant historical value for its information on the "Pioneer 440" is Jean Stansbury, *Bars on Her Shoulders: A Story of a Waac* (New York: Dodd, Mead & Company, 1943).

A good history of African Americans in the WACs is Martha S. Putney, *When the Nation Was in Need: Blacks in the Women's Army Corps During World War II* (Metuchen, N.J.: The Scarecrow Press, 1992). An important memoir by the highest-ranking black WAC of World War II is Charity Adams Earley, *One Woman's Army: A Black Officer Remembers the WAC* (College Station, Tex.: Texas A & M University Press, 1989). Janet Sims-Wood, Assistant Chief Librarian at the Moorland Spingarn Research Center at Howard University, Washington, D.C., is working on an oral history of black WACs and has published an excellent illustrated calendar of rare WAC photographs, *The World War II Black WAC Calendar, 1993* (Temple Hills, Md.: Afro Resources, 1993).

There is no book-length study of the WAVES. An overview of the World War II experiences of WAVES is contained in Susan H. Godson, "The Waves in World War II," *Naval Institute Proceedings* 107 (December 1981): 46–51. Godson is currently working on a book about the WAVES. Good information on the WAVES during World War II appears in Jean Ebbert and Marie-Beth Hall, *Crossed Currents: Navy Women from WW II to Tailhook* (Washington, D.C.: Brassey's, 1993). A useful review of literature about the WAVES is Regina T. Akers, "Female Naval Reservists During World War II: A Historiographical Essay," *Minerva* 8 (Summer 1990): 55–61. Akers is also working on a history of black WAVES during World War II. An unpublished administrative history is "History of the Women's Reserve" (Washington, D.C.: Naval Historical Center, 1946). A book of personal narratives is Marie Bennett Alsmeyer, *The Way of the WAVES: Women in the Navy* (Conway, Ark.: HAMBA Books, 1981). Also useful is Joy Bright Hancock, *Lady in the Navy: A Personal Reminiscence* (Annapolis: Md.: Naval Institute Press, 1972). A WAVE memoir published during the war years is Helen Hull Jacobs, *"By Your Leave, Sir": The Story of a Wave* (New York: Dodd, Mead & Company, 1943). A book that was written to introduce the work of the WAVES to the nation is Nancy Wilson Ross, *The WAVES: The Story of the Girls in Blue* (New York: Henry Holt and Company, 1943).

Very little has been published about the SPARs. A recent pamphlet containing useful historical information is *The Coast Guard & the Women's Reserve in World War II* (Washington, D.C.: Coast Guard Historian's Office, 1992). An unpublished, official history is "The Coast Guard at War, Women's Reserve" (Washington, D.C.: U.S. Coast Guard Headquarters, 1945). A book of personal narratives is Mary C. Lyne and Kay Arthur, *Three Years Behind the Mast: The Story of the United States Coast Guard SPARs* [Washington, D.C.: n.p., 1946].

A recently published, well-researched history of the Women Marines is Peter A. Soderbergh, *Women Marines: The World War II Era* (Westport,

Conn.: Praeger, 1992). An unpublished, official history was written by Pat Meid, "Marine Corps Women's Reserve in World War II" (Washington, D.C.: Historical Branch, U.S. Marines Corps, 1968). Two World War II memoirs are Neal Chapline, *Molly's Boots* (Detroit, Mich.: Harlo Press, 1983); and Frances Robinson Mitchell, *Experiencing the Depression and World War II* (Orono, Me.: Bear Paw Press, 1989). A book that was written to introduce the work of the Women Marines to the nation is Barbara A. White, *Lady Leatherneck* (New York: Dodd, Mead & Company, 1945).

Information on the Army and Navy Nurse Corps during World War II may be found in general histories of nursing, including Philip A. Kalisch and Beatrice J. Kalisch, *The Advance of American Nursing* (Boston: Little, Brown and Company, 1978); and Mary M. Roberts, *American Nursing: History and Interpretation* (New York: Macmillan Company, 1954). An important article on the efforts of Mabel K. Staupers to integrate black nurses into the military during World War II is Darlene Clark Hine, "Mabel K. Staupers and the Integration of Black Nurses into the Armed Forces," in John Hope Franklin and August Meier, eds., *Black Leaders of the Twentieth Century* (Urbana, Ill.: University of Illinois Press, 1982). A useful introduction to the history of the Army Nurse Corps during World War II is Judith Bellafaire, *The Army Nurse Corps in World War II* (Washington, D.C.: U.S. Army Center of Military History, 1993). Relevant information is also included in Robert V. Piemonte and Cindy Gurney, eds., *Highlights in the History of the Army Nurse Corps* (Washington, D.C.: U.S. Army Center of Military History, 1987). An unpublished, multivolume, official history of the ANC, which includes several volumes on World War II, is Pauline E. Maxwell, "History of the Army Nurse Corps, 1775–1948" (Washington, D.C.: U.S. Army Center of Military History, 1976). A memoir by a former Superintendent of the Army Nurse Corps during the war years is Julia O. Flikke, *Nurses in Action* (Philadelphia: J. B. Lippincott, 1943). Other ANC memoirs include Theresa Archard, *G.I. Nightingale: The Story of an Army Nurse* (New York: W. W. Norton, 1945); Juanita Redmond, *I Served on Bataan* (Philadelphia: J. B. Lippincott, 1943); and June Wandrey, *Bedpan Commando: The Story of a Combat Nurse During World War II* (Elmore, Oh.: Elmore Publishing Company, 1989). On the history of the Navy Nurse Corps, see *White Task Force: History of the Nurse Corps, United States Navy* (Washington, D.C.: Bureau of Medicine and Surgery, U.S. Navy, 1946). A personal history of the Navy Nurse Corps during World War II was written by Page Cooper, *Navy Nurse* (New York: Whittlesey House, 1946). For information on the Cadet Nurse Corps, consult *The United States Cadet Nurse Corps and Other Federal Nurse Training Programs* (Washington, D.C.: Government Printing Office, 1950).

The best general history of the Women Airforce Service Pilots is Marianne Verges, *On Silver Wings: The Women Airforce Service Pilots of World War II* (New York: Ballantine Books, 1991). Also useful is Sally Van Wegener Keil, *Those Wonderful Women in Their Flying Machines: The Unknown Heroines of World War II* (New York: Rawson, Wade, 1979). Jacqueline Cochran, the Director of the WASP, published a memoir, *The Stars at Noon* (Boston: Little, Brown and Company, 1954). Two recently published histories, written by former WASPs, are Jean Hascall Cole, *Women Pilots of World War II* (Salt Lake City: University of Utah Press, 1992); and Byrd Howell Granger, *On Final Approach: The Women Airforce Service Pilots of World War II* (Scottsdale, Ariz.: Falconer Publishing Company, 1991).

Good material on the work of the American Red Cross during World War II is included in Foster Rhea Dulles, *The American Red Cross: A History* (New York: Harper & Brothers, 1950). More specific information may be found in George Korson, *At His Side: The Story of the American Red Cross Overseas in World War II* (New York: Coward-McCann, 1945). A good account of the nurse recruiting efforts of the ARC during the Second World War is contained in Portia B. Kernodle, *The Red Cross Nurse in Action, 1882–1948* (New York: Harper & Brothers, 1949). An unpublished, multivolume, official history of the ARC, which includes at least ten volumes specifically on World War II, is "The History of the American National Red Cross" (Washington, D.C.: American National Red Cross, 1950). A memoir of an ARC worker in North Africa is Eleanor "Bumby" Stevenson and Pete Martin, *I Knew Your Soldier* (New York: Infantry Journal, Penguin Books, 1945). On the work of clubmobilers, see Marjorie Lee Morgan, ed., *The Clubmobile— The ARC in the Storm* (St. Petersburg, Fla.: Hazlett, 1982); and Oscar Whitelaw Rexford, *Battlestars & Doughnuts: World War II Clubmobile Experiences of Mary Metcalfe Rexford* (St. Louis, Mo.: Patrice Press, 1989).

For information about the work of the USO during World War II, see Julia M. H. Carson, *Home Away from Home: The Story of the USO* (New York: Harper & Brothers, 1946). A memoir describing the work of four well-known USO entertainers is Carole Landis, *Four Jills in a Jeep* (Cleveland and New York: World Publishing, 1944). Nancy Baird of Western Kentucky University is working on a book about the USO. On women correspondents during World War II, consult Julia Edwards, *Women of the World: The Great Foreign Correspondents* (Boston: Houghton Mifflin, 1988); and Lilya Wagner, *Women War Correspondents of World War II* (Westport, Conn.: Greenwood Press, 1989).

# Index

Adams, Marie, 24–25

African-American women: efforts to locate letters by, 4, 8; historical value of women's letters, 6; letters on military service of, 66–68, 68–71, 71–78; letter on social and economic hardships of in South, 90; opportunities for and segregation in military services, 65; prejudice and discrimination against in military service, 68; racism in postwar era, 255; support of community for World War II, 66

age: minimum requirements for women's services, 7, 39–40

Alaska: letters from, 21–22

Alexander, Paul, 40

Allen, Mariana Olweiler, 248–49

Alpha Kappa Alpha Sorority, 78

*American Journal of Nursing,* 20

American Red Cross (ARC). *See* Red Cross

American Volunteer Group ("Flying Tigers"), 202

American Women's Voluntary Services, 30

Anderson, Faye, 197, 258–59

Anderson, Margaret, 67

Army. *See* Army Nurse Corps (ANC); Women's Army Auxiliary Corps (WAAC); Women's Army Corps (WACs)

Army Nurse Corps (ANC): black nurses serving in, 65–66, 68; establishment of, 12; letters on racist practices in, 68–71. *See also* nurses and nursing

Atlantic Charter, 11–12

Australia: letters from, 180–81, 182–83

Avery, Dorcas C., 204

Band, Shirley, 40

Battle of the Atlantic, 23

Battle of the Bulge, 157, 173–74

*Battlestars and Doughnuts* (Rexford), 152

*Bedpan Commando: The Story of a Combat Nurse During World War II* (Wandrey), 125

Belgium: letters describing Battle of the Bulge, 157, 173–74

Berlin, Irving, 188–89

Berube, Allan, 62

Bethune, Mary McLeod, 71

Bolton Act (1943), 13

Boy, Anne Ferguson, 250–53

Brown, Dorothy, 170

Burma Road, 202. *See also* China-Burma-India (CBI) theater

Cadet Nurse Corps training program, 13, 30

Camp Nathan Bedford Forrest (Tennessee), 74

Camp Wheeler (Georgia), 83–84

Carey, Florence Jean, 144–45

Carney, Loretta, 60–61

censorship: letter from European front on, 152–53; letter on process of in Pacific war, 183; strictness of regulations and information provided by letters, 6

*Chicago Defender:* letters to on racism in military, 75–76

China: Japanese occupation of, 201–2; letters from women serving in during Pacific war, 210–18; Red Cross volunteers during Japanese invasion of, 11

China-Burma-India (CBI) theater: service of women in, 178, 201–2

Civil War, American, 12

clubmobiles: Red Cross on European front, 152–55

Coast Guard: establishment of women's service, 29–30; letters by men objecting to military service of women, 40. *See also* SPARs

Cockram, Beatrice, 161–63

Cofer, Maud Turner, 171

combat nurses: D-Day and invasion of France, 158–60; European front during summer and autumn of 1944, 163–68, 171–76; in Italian Campaign, 127–29, 131–36; in North Africa, 125–27, 130

*Coming Out Under Fire: The History of Gay Men and Women in World War Two* (Berube), 62

concentration camps: letters describing, 223, 225–26; tasks facing uniformed personnel at end of World War II, 221. *See also* prison camps

Conter, Monica, 17–19

Coolen, Mary B., 41

Crowell, Shirley Magowan, 44–46

Dachau concentration camp, 225–26

D-Day: arrival of Army nurses and Red Cross hospital workers in France, 121, 158; letter describing stateside response to, 98; Red Cross workers and preparations for, 155–57

"Dear Jane" letters, 141–42

demobilization: logistics of, 245–47; point system of, 219, 247

dietitians: letters from evacuation hospital in Italy, 136; letter from

Veterans Administration hospital in U.S., 60

Donahue, Betty Oliver, 189–94

"Double V" campaign, 66

"doughnut girls": Red Cross on European front, 151

Dykstra, Matilda E., 203

Eaton, Mary, 160–61

education: letters discussing postwar plans for, 92, 112, 247–48, 248–49; minimum requirements for women's services, 7

employment: opportunities for women in postwar era, 255; possibilities for women in military, 50

England. *See* United Kingdom

entomology: wartime research by women on mosquitos and tropical diseases, 145–46

Europe: refugees after end of war in, 228–29; service women stationed in during war against Axis, 121–22; uniformed women as back-up support for invasion forces, 170–71. *See also* France; Germany; United Kingdom

experience: letters on meaning and significance of wartime, 256–60

Faneuf, Marjorie LaPalme, 171–76

Farris, Betty Ann, 52–54

Fleming, Alexander, 133

flight orderly: letters describing work as, 52–54, 245–47

"Flying Tigers" (American Volunteer Group), 202

Forsyth, Clementine B., 79

France: arrival of Army nurses and Red Cross workers following D-Day, 121; letters describing D-Day and invasion of Normandy, 158–60; letters describing conditions in summer and autumn of 1944, 160–61, 161–63, 163–68; letter describing wedding on European front, 169–70

Freid, Bernice Sains, 63–65

Gellhorn, Martha, 158
Germany, letters from: combat nurse on European front, 175–76; on coping with death of war casualties, 226–28; describing concentration camps, 225; describing end of war, 223–24; on experiences of WAC in Army of Occupation, 235–38. *See also* Europe; World War II
Giles, Helen McKee, 130–36
Girarde, Lillian C., 21–22
Goode, Annabel, 222–23
Gordon, Linda, 8
Great Britain. *See* United Kingdom
Groppe, Major Edna B., 70–71
Guam: capture of Navy nurses by Japanese, 177; letters on postwar aid to war victims, 229–32

Halverson, Pearl Gullickson, 102–8
Harden, Frances, 185–89
Haring, Jacqueline, 194–97, 238–43
Harris, Mary C., 67
Hart, Barbara Drake, 204–10
Hastie, William H., 66
Hawaii, letters from: describing hospital and nursing work during Pacific war, 179–80; at start of war, 13–19; Women Marines during Pacific war, 218–19. *See also* Pearl Harbor
Hess, Ruth, 158–60
Hicks, Dorothy, 20–21
Hiroshima: bombing of and prospects for postwar world, 222, 223; letter describing after end of war, 241–42
Hitler, Adolf: letter discussing death of, 189
Hobby, Col. Oveta Culp, 87
Hodgson, Marion Stegeman, 114–20
Hogan, Aileen, 163–68
Holcomb, Gen. Thomas, 42
Holiday, Nellie R., 77–78
Holtman, Florence, 143–44
homosexuality. *See* lesbians and lesbianism
hospitals: letters from evacuation hospital in Italy, 136–43; letters describing life on hospital ship in Pacific, 200–201; Red Cross support services in, 59. *See also* nurses and nursing
hospital ships: Navy nurses serving in Pacific war, 199–201
Hudson, Miriam Dawson, 253–55

Iceland: letters from, 22–23
India: letters from, 202–10
Iran: letters from, 144–45
Italy: arrival of Army nurses in, 121; fall of Rome in 1944, 143; letters from evacuation hospital in, 136–43; letters describing Italian Campaign, 127–29, 131–36; letters from Red Cross club director, 143–44

Japan: letters from describing service in Army of Occupation, 239, 241–43, 244–45; service of WACs in Army of Occupation, 235; service of women in war against, 177–78. *See also* China-Burma-India theater; Pearl Harbor; prison camps; World War II
Jews: historical value of women's letters, 6; letter describing service in military, 64–65
John, Betty B., 49–50, 146–50
Johnson, Dovey M., 71–72
Johnson, Thomasina Walker, 78–79

Katz, Rosie, 42–44
Kiernan, Blanche M., 13–17, 179–80
Kinsey, Margaret Combs, 35–38
Kintzel, June, 245–48
Knigin, Tamaranth D., 145–46

Leavitt, Hester J., 150–51
Lend Lease Act (1941), 11
lesbians and lesbianism: appeal for tolerance and understanding for in postwar era, 255–56; historical value of women's letters, 6; letter on problems of in military, 62–63
*Libby* (John), 150

Los Banos Internment Camp, 177
Louisiana maneuvers of September
    1941, 20–21
Lundberg, Carolyn B., 41


Maas, Erna Elizabeth, 226–28
MacArthur, Gen. Douglas A., 180
McDermott, Eunice McConnell, 93–
    102
McDonald, Dorothy, 259–60
March on Washington Movement, 66
Marines: establishment of women's
    service, 30, 41–42. *See also* Women
    Marines
marriage: letters on postwar plans,
    250–53; letter describing wedding
    on European front, 169–70; letter
    describing wedding in New Guinea,
    187–88; regulations concerning
    women in military services, 61–62
Marshall, Gen. George C., 31
Matthews, Ann F., 229
Mauldin, Bill, 133
mechanics, aviation: letters describing
    service as, 51–52, 218–19
Metcalfe, Mary, 151–55, 223–24
military: African-American women and
    service in, 65, 66; employment
    possibilities for women in, 50;
    letters describing jobs held by
    women in, 50–52; minimum age
    requirements for service in, 7, 39–
    40; orientation programs for newly
    commissioned nurses, 46–49;
    overview of service of women
    during World War II, 29–30; social
    lives of women serving in, 61. *See
    also* specific branches of service
Minton, Madge Rutherford, 54–57
*Miss You: The World War II Letters
    of Barbara Wooddall Taylor
    and Charles E. Taylor* (Litoff &
    Smith), 3
Mitchick, Evelyn Zimmerman, 169–70
Mixsell, Mary, 180–81
Moor, Emogene Heston, 235–38
Morehouse, Charlotte, 30–35

Nagasaki: bombing of and prospects
    for postwar world, 222, 223
National Association of Colored
    Graduate Nurses (NACGN), 68
Native Americans: service of women
    in military during World War II,
    170
Naval Air Evacuation Service: Navy
    flight nurses, 199–200
Navy. *See* Navy Nurse Corps;
    WAVES
Navy Nurse Corps (NNC): black
    nurses serving in, 66; establishment
    of, 12; hospital ships in Pacific war,
    199–201. *See also* nurses and
    nursing
Necko, Mary Cugini, 108–13
Nelson, Constance E., 74–75
Netherlands: letters from, 172–73,
    174–75
New Guinea: letters from, 184–93,
    194–97
*Newfoundland* (ship), 129
North Africa: arrival of Army nurses
    in, 121; letters describing campaign
    in, 123–25, 125–27, 130
nuclear war: prospects for postwar
    world, 222, 223
Nuremberg trial, 238
nurses and nursing: black nurses in
    military services, 65–66; on hospital
    ships in Pacific war, 178; killed in
    line of duty on European front, 168;
    military orientation programs for
    newly commissioned, 46; minimum
    educational requirements for
    women's services, 7; Naval Air
    Evacuation Service, 199–200;
    overview of military service during
    World War II, 12–13; as prisoners
    of war in Pacific, 177; support roles
    of in European invasion, 171. *See
    also* Army Nurse Corps; combat
    nurses; hospitals; Navy Nurse Corps
nurses, letters: from Australia, 180–81;
    from black nurses on discriminatory
    practices in military, 68–71; on
    capture of ship by Germans, 25–28;

on D-Day and invasion of Normandy, 158–60; on death of war casualties, 226–28; from European front in 1944, 171–76; on German concentration camps, 225–26; from Hawaii during Pacific war, 179–80; from Hawaii at start of war, 13–19; from hospital ships during Pacific war, 200–201; from India, 202–4; on London Blitz, 23–24; on Louisiana maneuvers of September 1941, 20–21; on military orientation programs, 46–49; from Italy, 127–29, 131–36; from North Africa, 125–27, 130; from Northern Ireland, 122–23; on postwar care for war victims in Pacific, 229–32; on refugees in Europe, 228–29; on release of prisoners of war from Japanese internment camps, 232–35; from remote outposts at start of war, 21–23

occupations: of women in military, 50. *See also* employment
officers: status of nurses as, 12
Oiness, Sylvia, 25–28

Pateman, Yvonne, 58
Pearl Harbor, Japanese attack on: care of wounded by Army and Navy nurses and Red Cross workers, 177; entry of U.S. into World War II, 11, 12; letters describing, 16–17, 17–19
penicillin: letters discussing use of, 131, 133
Pennington, Clyde, 130
Philippines, letters from: after end of war, 239–41, 243–44; during Pacific war, 193–94, 197–98, 198–99; at start of war, 24–25
Piatt, Margaret Anderson, 136–43
Pilkey, Rita, 210–18
pilots: letters from women serving as, 54–57, 58–59, 114–20. *See also* Women Airforce Service Pilots (WASPs)

Pope, Frances C., 124
postal workers: WACs in England as, 171
postwar period: employment opportunities for women, 255; letters discussing plans for education, 92, 112, 247–48, 248–29; letters discussing plans for marriage, 250–53; letters on hopes for children, 253–55; letters on meaning and significance of wartime experiences, 256–60
prison camps: letters describing experiences in Japanese internment camps in Philippines, 25, 177; letters on release of prisoners of war from Japanese internment camps in China, 232–35. *See also* concentration camps
prisoners of war: Army and Navy nurses in Pacific war as, 177; letter on black Army nurses and care of Germans, 69; letter describing German attack on shipping and experience as, 25–28; letter describing German prisoners of war on European front, 166–67; letters on release of from Japanese internment camps, 232–35; letter on repatriation of Americans during Pacific war, 196; tasks facing uniformed personnel in Europe at end of World War II, 221

quotas: for blacks in military services, 65, 66

Raab, Harriet, 59–60
racism: letters from black military nurses on, 68–71; letters from black women serving in WACs, 71–78; in postwar era, 255
Radioman: letter from Women Marines on service as, 42–43
Randolph, A. Philip, 66
Red Cross: minimum age and education requirements for service in, 7; service in Europe during war

Red-Cross (*continued*)
   against Axis, 121, 122, 171; war
   casualties and hospital support
   services of, 59; war-related uniforms
   worn by women in, 30
Red Cross, letters: from China during
   Pacific war, 210–18; from club
   director in England, 150–51; from
   club director in Italy, 143–44; on
   conditions in France in summer and
   autumn of 1944, 161–63; describing
   bombing of hospital ship, 129–30;
   from "doughnut girl" on European
   front, 151–55; from India during
   Pacific war, 204–10; on London
   Blitz, 23–24; on meaning and
   significance of wartime experiences,
   256–60; on military orientation
   programs, 49–50; from New Guinea
   during Pacific war, 194–97; from
   Philippines after end of war, 239–
   41; from Philippines during Pacific
   war, 197–98; from Philippines at
   start of war, 24–25; on preparations
   for D-Day in United Kingdom,
   155–57; on service clubs in Great
   Britain, 146–50; on snack bars in
   Iran, 144–45
*Refuge* (hospital ship): letter
   describing activities aboard,
   201
refugees: letter on European after end
   of war, 228–29; letter on North
   African, 130; tasks facing uniformed
   personnel at end of World War II,
   221
Rexford, Mary Metcalfe, 151–55,
   223–24
Reynolds, Georgia, 200
Rivers, Beatrice, 232
Rogers, Edith Nourse, 29
Rome: fall of in 1944, 143
Roosevelt, Franklin D.: letter
   commenting on election of, 89;
   letters on death of, 90–91, 175;
   letter describing funeral of,
   112

Salvation Army, 30
Santo Tomas Internment Camp, 25,
   177
Schoonover, Lillian, 229
scientists, women as: entomology
   research during World War II,
   145–46
Selective Service and Training Act
   (1940), 11
service clubs: Red Cross in Great
   Britain, 146–50
shipping: letters describing German
   attacks on, 25–28, 129. *See also*
   hospital ships
Simon, Gysella, 155–57, 256–58
"slander" campaign: against WAACs,
   33, 35
Slanger, Frances Y., 168–69
Smith, Margaret Chase, 29, 104, 105,
   178
social lives: of women in military
   services, 61. *See also* marriage
social workers: letters from Red Cross,
   24–25
Spanish Civil War, 11
SPARs: acceptance of African
   Americans into, 65, 78; duty in
   Hawaii and Alaska during Pacific
   war, 178; establishment of, 29–30
SPARs, letters: describing stateside
   duty, 102–108; on minimum age
   requirements for service in, 39–40;
   on name of, 38–39; on privilege of
   serving as officer, 41
Staupers, Mabel, 68
Stratton, Capt. Dorothy C., 38–39,
   104
Streeter, Ruth Cheney, 42


Taylor, Barbara Wooddall, 3
Telford, Peggy, 129–30
Tibbs, Pearl E., 67–68
travel: discrimination against black
   service personnel on public
   transportation, 72–74; letter
   describing wartime conditions of,

89; service in military and
opportunities for, 63–64
Trickey, Katherine, 82–93
Truman, Harry S.: end of segregation
in military, 255; proclamation of
end of war, 221
Turnbull, Lorraine, 50–52, 218–19

United Kingdom, letters from: Army
nurses in Northern Ireland, 122–23;
during London Blitz, 23–24; Red
Cross club director, 150–51; Red
Cross workers preparing for
Normandy invasion, 153–55, 155–
57
United Nations Relief and
Rehabilitation Agency (UNRRA),
221, 228–29
United Service Organizations (USO),
30

V-E Day: reactions to announcement
of, 175–76, 221
V-Mail: definition of, 126n.5
"V-Mail Update" (newsletter), 4
veterans: status of WASPs as, 54, 120

Waesche, Adm. Russell R., 104
Wandrey, June, 125–29, 225–26
WAVES (Women Accepted for
Volunteer Emergency Service):
acceptance of African Americans
into, 65, 78–79; duty in Hawaii and
Alaska during Pacific war, 178;
establishment of, 29
WAVES, letters: describing
demobilization process, 245–47;
describing stateside duty, 93–102;
on experiences of Jewish woman in,
64–65; on postwar plans, 248–49;
on travel in, 63–64; on visits of
dignitaries, 35–38; on work as flight
orderly, 52–54
Wayman, Martha Alice, 181–85, 243–
45
Wesley, Carter, 74

West, Harriet M., 71, 72–73
Wetherby, Mary Ann, 58
Women Airforce Service Pilots
(WASPs): establishment of, 30;
military status of women serving in,
54, 120; refusal to accept African
Americans, 65; requirements for
entry into program, 7
Women Airforce Service Pilots
(WASPs), letters: describing
stateside duty, 114–20; on love of
flying, 54–57, 58–59
Women Marines: duty in Alaska and
Hawaii during Pacific war, 178;
establishment of, 30; refusal to
accept African Americans, 65
Women Marines, letters: describing
stateside duty, 108–13; on
employment as aviation mechanic,
51–52; from Hawaii during Pacific
war, 218–19; on training and
orientation, 42–46
Women's Army Auxiliary Corps
(WAAC): establishment of, 29;
treatment of African-American
women serving in, 65
Women's Army Auxiliary Corps
(WAAC), letters: on Army of
Occupation in Germany, 235–38;
describing service in North Africa,
123–25; on training and orientation,
30–35
Women's Army Corps (WACs):
establishment of, 29; numbers of
stationed in Europe, 121; racial
segregation and discrimination in,
71; service in Pacific war, 178;
support roles in European invasion
forces, 170–71
Women's Army Corps (WACs),
letters: from African-American
women serving in, 71–78;
describing life in Southwest Pacific,
181–85, 185–89, 189–94, 198–99;
on service in Army of Occupation
in Japan, 243, 244–45; from women
on stateside duty, 82–93

Women's Auxiliary Ferrying Squadron (WAFS), 30

Women's Flying Training Detachment (WFTD), 30

women's history: historical value of women's letters from World War II, 6–7, 9

Women's Land Army, 30

World War II: China-Burma-India theater, 201–18; efforts to locate letters, 3–9; entry of U.S. into, 11–12; end of and prospects for postwar world, 91, 221–22; overview of service of women in military during, 29–30; service of women in continental U.S. during, 81; service of women in Europe during, 121–22; service of women in war against Japan, 177–78. *See also* specific topics